T0296513

The Royal Society of London, effectively Britain's national academy of science, has, ever since its foundation in 1660, been particularly concerned with experimental science. Early members like Boyle, Hooke and Newton all advocated and practised experiment. The Society itself for long filled its weekly meetings with the performance of experiments together with accounts of experiments performed elsewhere. This book for the first time explores the practice of the Society's members and shows how this altered between 1660 and 1727.

Promoting Experimental Learning

Promoting
Experimental
Learning

Experiment and the Royal Society
1660–1727

MARIE BOAS HALL

The right of the
University of Cambridge
to print and sell
all manner of books
was granted by
Henry VIII in 1534.
The University has printed
and published continuously
since 1584.

CAMBRIDGE UNIVERSITY PRESS
CAMBRIDGE
NEW YORK   PORT CHESTER
MELBOURNE   SYDNEY

Published by the Press Syndicate of the University of Cambridge
The Pitt Building, Trumpington Street, Cambridge CB2 1RP
40 West 20th Street, New York, NY 10011–4211, USA
10 Stamford Road, Oakleigh, Melbourne 3166, Australia

First published 1991

Printed in Great Brtiain at the University Press, Cambridge

British Library cataloguing in publication data available

Library of Congress cataloguing in publication data available

ISBN 0 521 40503 3 hardback

To my former students
who have so kindly remembered me

Upon a Review of the whole, this Experiment will possibly not only Surprize and Amuse some, but Please and Delight others; and not only so, but perhaps afford some Instruction to a Philosophical Genius.

F. Slare, *Phil. Trans.*, 18 (1694), 213

Contents

Figures

Preface

The roots of this study lie deep in the past: I have been interested in the history of the early Royal Society ever since, in 1951, I was generously allowed by the Society's then President, Council and Librarians to plunge into a study of the Boyle papers, the first to wish to do so, probably, since the mid-eighteenth century. This interest has grown with the years, especially when, some ten years later, my husband and I began to edit the correspondence of Henry Oldenburg, named as one of the two Secretaries in the Charters of 1662 and 1663. In those days historical interest in the Royal Society centred mainly on the Society's origins, while currently it is rather on its institutional aspects and its sociological roots. Very little attention has been given, even now, to exactly what went on at its meetings, and especially what its Curators and Operators did in return for their salaries and what they contributed to its meetings, then its main activity. In the twentieth century, as in the seventeenth and eighteenth, many, perhaps most, Fellows value the Society principally for the honour which election to it confers, only a minority taking an active rôle in its administrative affairs. But for those seventeenth-century Fellows who best exemplified the Society's aims, it was participation in the meetings which counted, and hence what went on at those meetings reveals much about their interests and those of the Society as a whole. For reasons which will become apparent below, I chose to concentrate upon one aspect of the Society's work, namely its concern with experiment. I leave to others various other concerns of the Society of the early years: the history of trades, medicine, astronomy, mathematics, and so on.

I began by extracting from Thomas Birch's invaluable *History of the Royal Society* information and statistics about the density of experiment up to 1687 and followed this by doing the same for the Journal Books for the years after 1687. This I found a laborious task, but one not without interest or reward. It is, of course, difficult or rather impossible to be accurately quantitative about these things. What precisely constitutes an experiment is not easy to determine, and what figure can or should one assign to 'several experiments' performed by one individual? And what about the showing of models or instruments? Hence I am all too aware that my statistics lack the limpidity and precision that would be desirable; I can only say that I have done my best.

When I began this work I was not aware of John Heilbron's excellent and painstaking study *Physics at the Royal Society during Newton's Presidency* (1983), or perhaps I should not have been brave enough to begin on mine, on which I was well embarked when I did come across Heilbron's monograph, at the end of 1987. It should be obvious that the many papers by Michael Hunter, of which he was kind enough to send me offprints and which are incorporated in his 1989 *Establishing the New Science* (of which he also generously sent me a copy), appeared during the course of my work. His earlier monograph *The Royal Society and its Fellows 1660–1700* was immensely useful to me as it must be to all students of the earlier Royal Society. He also kindly read an early draft of my work and made numerous suggestions, from which I have benefitted. A most useful aid was the name index to Birch's *History* prepared by Gail Edwald Scala under the supervision of Dorothy M. Schullian, formerly Curator of the History of Science Collections of the Cornell University Library, who very thoughtfully sent my husband and myself an offprint (from *Notes and Records of the Royal Society*). I should also like to thank Paolo Casini and Simon Schaffer for useful offprints.

My greatest thanks must be to the ever-helpful librarians of the Royal Society who have, during nearly forty years, patiently provided much tolerant assistance as well as friendship. I also thank warmly the Presidents and Councils of the Royal Society who have over the years given me permission to read in the library, to reproduce manuscripts from the Society's archives, and to feel at home within its walls.

My particular thanks, as always, go to my husband for his support, both professional and personal. And also to Isaac Newton Felis for his constant and (usually) patient companionship.

Marie Boas Hall

Tackley, Oxford
1985–1990

A note on dates

As is normal, I have effected a compromise between the English Old Style dates and the Continental acceptance of the Gregorian Calendar (New Style, N.S.). Dates here are Old Style (O.S.) as regards days (then ten days behind New Style dates) but quasi-New Style as regards years, that is beginning the year in January, not in March, but, as the more up to date English commonly did, writing both years for the first three months of the year: thus the first meetings which led to the Royal Society occurred in November and December 1660, and continued in January 1660/61. As far as Royal Society activities are concerned, I have modernised in another way. Logically, for meetings after 1662, I should have begun the year with St Andrew's Day (30 November) when from 1663 onwards elections were made to the Council and officers elected. But, as this would have resulted in a discontinuity between the first three years and subsequent years, and as it would have seemed strange to all readers not closely connected with the modern Royal Society, I have chosen in the interests of clarity and simplicity to use calendar years throughout, even though this may slightly prejudice the statistics in the figures showing experimental activity.

~ 1 ~

Introductory

> The business of the Society in their Ordinary Meetings shall be to order, take account, consider, and discourse of philosophical experiments and observations; to read, hear, and discourse upon letters, reports, and other papers concerning philosophical matters; as also to view, and discourse upon, rareties of nature and art; and thereupon to consider, what may be deduced from them, or any of them; and how far they or any of them, may be improved for use or discovery.

So run statutes of the Royal Society as passed in 1663 and so also those of 1939 and ever since. Moreover, in 1663 there was also a separate chapter of the statutes devoted to the making of, reporting on and financing of experiments.[1] But the statutes of 1847, which were intended to embody new reforms which should render the Royal Society more strictly scientific than it had become in earlier decades, read, starkly,

> The business of the Society in their ordinary Meetings, shall be to read and hear letters, reports, and other papers, concerning Philosophical matters.

These mid-nineteenth-century statutes in fact recognised a situation which had existed throughout much of the previous hundred years, a period during which both experiment and discussion were slowly abandoned and the character of the Royal Society's meetings altered, changing from an atmosphere of lively discussion and debate and the frequent display of experiment to one which was determinedly formal and lifeless.

So, paradoxically, the Royal Society no longer reflected the practice of contemporary science which was certainly devoted to experiment. In 1839, and again nine years later, the physiologist Marshall Hall was to complain bitterly at the rejection of his papers by the Royal Society's Physiological Committee.[2] The burden of his complaint was that members of the Committee either misunderstood or disbelieved his claims for experimental proof of his discoveries and interpretation, and, worst of all, refused to let him show them the experiments he was interpreting. Yet the original 'designe' of the Royal Society, as its Statutes confirm, had been 'the promoting of Physico-Mathematicall Experimental Learning', which by no means excluded biological experiment, and in the seventeenth century it had been expressly stated that not only should its weekly meetings have for their purpose 'to consult and debate concerning experimental learning' of all kinds, but to perform experiments at the meetings. As a Fellow could say in the late 1660s, 'The businesse of the Society is to make experiments',[3] yet by the early nineteenth century meetings were purely concerned with the reading of papers, which might be based upon experiment and even describe experiments performed, but which were never accompanied by experimental demonstration.

This was, obviously, not because the importance of experimental science had been lost sight of, for no one could describe nineteenth-century scientists as opposed to experiment. Nor was it that the Fellows were no longer good experimentalists (the regular award of the Copley Medal specifically for experimental science excludes this hypothesis), nor that nineteenth-century scientific experiment was unsuited to public demonstration (the dramatic experiments regularly performed at the Royal Institution lectures by Davy and Faraday show the contrary). Clearly, the conventions of the Royal Society meetings had changed in some way while their purpose had changed as well.

To discover how these changes came about it is necessary to trace the conduct of the meetings and the general activities of the Society from its inception into the eighteenth century; this will also give some precise idea of what the Royal Society actually did during its first three-quarters of a century. This is a task not previously fully attempted, although a number of studies in the past few years have examined particular aspects of the question. The most closely related to what is attempted here is Heilbron's analysis of *physical* experiment at the Society during Newton's Presidency (1703–27), a study which within the compass of its subject is very thorough, analysing both the content of the meetings and the publication of experimental papers in the *Philosophical Transactions* (by no means the same thing, since that journal was not yet the official organ of the Society).[4] This is an excellent work, with the one flaw that it takes for granted that the only purpose of the Society was the advance of natural science (not learning in general) and physical science at that. A good many years ago now I discussed the way in which certain leading

Fellows of the early Royal Society contributed to the advance of experimental philosophy, concluding that their private contributions were often more important than their public ones, since they tended to experiment elsewhere than at the Society's rooms to advance their main work; this study was not very inclusive and was limited to the period before about 1680.[5] More recently, Hunter has discussed the institutional workings of the Society, while Hunter and Wood have published various seventeenth-century proposals for the reform of the Society (all of which stress the importance of experiment). Hunter's work supplements the many discussions by historians of science of the original aims of the Society as manifested in its charter, but does not tell us what was actually done by the Society at its meetings; indeed, much of what Hunter discusses is, ironically, a record of failure.[6] More recently still, Shapin and others have surveyed experimental activities in England inside and outside the Society during its early years from a sociological point of view, with particular emphasis upon the location of experimental work and its public or private nature, especially during the 1660s and 1670s.[7]

However, no one so far has analysed the whole of the Society's day to day activities as they relate to experiment, clearly revealed in the Journal Books, not just for its first fifteen years of existence – admittedly some of its most important – but for the less familiar later years when the founders were aging or dead, the first enthusiasm had dampened, and, as the members themselves were aware, new methods for the conduct of meetings were needed. What the Society really did during the later decades of the seventeenth century has been little studied in recent years. To round off an investigation of this kind it has seemed advisable to include a briefer survey of the Society's activities as they relate to experiment during the first quarter of the eighteenth century, chiefly the period of Newton's Presidency.

As all students of the early Royal Society are keenly aware, experimental activity at meetings after November 1663 depended very much for much of the century upon the work of its first Curator of Experiments, Robert Hooke. Hooke was an extremely able and inventive experimenter who gave experiment a very high place in his view of proper scientific method. It might therefore have been expected that when he became Secretary in 1677 and played a more dominant rôle in the Society's affairs, experiment in turn would have had a greater rôle in the meetings, in accordance with his frequent declarations of its importance for natural philosophy. The record of the Journal Books shows this not to have been the case. Hooke, like many others, was often happier to talk about experiments already performed or to suggest experiments that might be performed than to go through the time-consuming, costly and laborious business of performing experiments suitable to the topic in hand. And, again like many others, in old age his interests tended to turn from experimental to descriptive science and even to such subjects as mythology or archaeology, not directly connected with the study of nature as usually conceived. Newton, of course, also interested himself in what, to moderns, are peripheral subjects, such as biblical exegesis and chronology, but until the last years of his long life he was actively concerned with experimentation and effectively saw to it that the Royal Society was as well. In this connection it should be recalled that much brilliant and important work in the biological sciences was presented to meetings during the later years of his Presidency.

It has proved possible to show how discussion gradually dwindled and died at the Royal Society's meetings as the reading of papers replaced in very large part spontaneous discussion and the performance of experiment (which always provoked it), but during the period under discussion here comment upon the papers read was always welcomed, even if not always given. It has not proved possible to discover the reason why discussion of any sort disappeared so thoroughly as to be taken by the later eighteenth century to be counter to the statutes. It is only possible to observe that it did so. That the performance of experiment came to such a complete stop is equally surprising. In part this may be laid at the door of the last effective Curator of Experiments, J. T. Desaguliers, who performed fewer and fewer experiments once Newton ceased to direct him in them. He did occasionally perform experiments as late as the 1730s, mainly apparently in order to claim the reward of money from Copley's legacy (before it paid for a medal). Others too continued to produce experiments, but these became fewer and fewer as time went on. To this extent, the common historical opinion that the Royal Society was in decline in the early and mid-eighteenth century is justified.[8] But for a rounded picture it is necessary to judge the Society not only in terms of the conduct of its meetings as revealed in the minutes but also in terms of the standing of its Fellows and its general reputation. Hence it has seemed right here to include some discussion of the aims that the Society held as seen by the members

(Chapter 2), of the means by which the attitude of the Society was disseminated throughout the world (Chapters 4, 6 and 8), and of the way in which the Society was viewed by contemporaries (Chapter 9). But the main emphasis here must be on the detailed record of the minutes and an analysis of the place experiment held in the regular conduct of the meetings.

To assist the reader in understanding this analysis, I have, in accordance with modern historical usage, provided graphs to illustrate experimental activity in the three periods into which this study is divided. These attempt to show both the approximate number of experiments performed annually at meetings and, for the period after 1674, the number of experimental papers, discourses or letters read at meetings during each year. Clearly any such analysis must be highly subjective, and I make no claims to precise accuracy; yet I hope that my graphs will be a useful guide and show at least relative activity if they cannot show absolute numbers. They should provide some help in reading the verbal analysis which presents the facts and events upon which the figures are based. For convenience, I have also provided three lists (Figures 1, 5 and 6) which give the names of the Officers and principal employees during the period 1662 to 1727.

In both the graphs and the text I have tried to distinguish roughly between *experimental* learning and *observation*, but without attempting any exact definition of experiment. I have not counted the presentation of random empirical facts, fancies or thoughts as experiment, nor in consequence papers based upon them as experimental. I have not counted histories of trades, though these were clearly empirical. Nor have I counted the display of instruments or models as experiment, except where this involved their use for the presentation of experiment. Obviously I could have done these things and some will argue that I should have done. Seventeenth-century usage might seem to justify it, but I chose to take a slightly more rigorous attitude for reasons which will appear below. My justification for this is not mere simplicity, and it is, above all things, not for the passing of judgement. It is rather that, as one must always keep in mind, the Royal Society appointed Curators of *Experiment* (whether long or short term) and not Curators of Observation, of Empirical Record nor of Instruments. The Society of course welcomed the presentation of observational accounts or descriptions of instruments or the showing of models. And it is true that the paid Curators of Experiment, like Hooke, seem sometimes to have regarded reports of observations made with instruments or description or models of instruments displayed as being at least partly equivalent to the showing of experiment, especially when the instrument itself was available for demonstration. Yet one often gets the impression that this was 'fudging', that is, that the Curators knew that they were in such cases not so much presenting an experiment as presenting a substitute or preliminary to experiment. Experiments, after all, were often difficult or tedious to present; one intended for a meeting might not work or might not be suitable for presentation to an audience. In such cases instruments and models were an acceptable and easy substitute and always attracted interest. It is true that, obviously, this all makes strict quantitative analysis impossible. Yet it does not negate the possibility of relative quantitative analysis such as is displayed in the graphs, where it is clear that in some years more experiments were presented to meetings than in others. And this is all that I have aimed at, endeavouring to present the material with as much consistency as is reasonably attainable. *Empiricism* by itself is too vague a term to permit any quantitative or qualitative analysis, since it includes hearsay, sense impressions, supernatural experience and statistics as well as genuine observation. Unless some restriction is to be made, the whole exercise is meaningless. Any reader who finds this too restrictive must make allowances. I do not believe that the absence of such empirical presentations seriously falsifies my general conclusions in any way. And on the whole it is fully in keeping with seventeenth- and early eighteenth-century usage largely to abstract the performance of experiment from other forms of empirical presentation, not to suggest that it was better, but to understand the genuine role of experimental learning. That this is as valid for the early eighteenth century as it was for the later seventeenth century is clear from the terms of the Copley Award (after 1731 the Copley Medal) which explicitly specify *experimental* work only as admissable for the award.

Thus I surely need make no apology for abstracting the performance of experiment from other activities at meetings, while always noting the reading of experimental papers. Such activities as the reading of papers and letters not concerned with experiment, conversation, random comment and recollection, general discussion and so on, although informative and often including observation, must therefore be excluded. It should be obvious that this must have been only a part of what went on at meetings, for not all the proper concerns of the Royal

Society (that is, what was thought proper by the Fellows) by any means lent themselves to experiment. Mathematical learning clearly came within the Society's interests, but the reading of pure mathematics could hardly be popular and was seldom attempted. Papers on applied mathematics, if short, were sometimes read, but not often; publication in the *Philosophical Transactions*, when that was available, was usually deemed sufficient, sometimes after brief consideration. (It is a measure of the Fellow's extraordinary appreciation of Newton's *De motu corporum* that it was read at meetings.) Astronomical papers were read not infrequently when theoretical, but papers of astronomical observations were 'received' (that is, mentioned or described to a meeting) and then recommended for printing. Meteorological observations when detailed were similarly treated, although theoretical comments might be read. The same considerations applied to geological and medical reports. All these subjects might produce papers of a high empirical content, but equally might be mathematical or theoretical. Chemistry was clearly experimental. Alchemy, which appeared infrequently, normally was presented only when empirical rather than theoretical; even theoretical chemistry was avoided as inclining towards 'system building' and requiring a judgement of theory which the Society regarded itself as pledged not to give.

I do not intend anywhere to offer value judgements. When I say that experimental activity languished I mean just that: that there was much less of it than had been the case in other years and than the Fellows generally wished there to be. I do not mean to imply that experimental activity was in some way 'better' than other activity (in spite of the fact that some Fellows at every period under discussion did say so). In quantifying experimental activity I do not at all intend to suggest that only experimental activity was proper for presentation, although, as will become amply clear below, it was a repeated complaint of the Officers, Councils and many Fellows that experimental activity ought to be maintained at a high level, and they did not say this about the presentation of instruments, models or histories of trades. This is clearly shown by the way that Curators of Experiment were sought for and even paid out of the Society's meagre funds. The Society was prepared to go to considerable lengths to ensure the presentation of experiment and experimental papers at meetings. And, as I have tried to show in Chapter 2, most of the plans for reorganising the Society in this period revolved around increased experimental activity, while (Chapter 9) critics outside the Society, both at home and abroad, judged the Society by such activity. Foreign savants regarded such activity as the chief characteristic of the Society and what differentiated it from other learned bodies (exceptions were the short-lived Accademia del Cimento and the also short-lived Académie de Physique of Caen). This was such a universal opinion that I commit no solecism in choosing to explore only the experimental content of meetings.

This I have tried to do without regard to subject, not, as Heilbron ably did for the Newtonian period, concentrating on physical science, but endeavouring to consider all subjects and all experimental learning. Once again I must stress that I intend no value judgements, conscious of the views on this point of modern historians. But I cannot resist pointing out that this is to be ahistorical to the extent that the period I am considering was one in which value judgements were constantly made, being considered moral and necessary to a rightly ordered society. I have here attempted only to present as clearly as possible what the Fellows thought about the uses of experiment, what they did to promote experiment, especially its presentation at their meetings, and how the quantity of experiment changed over the years between 1660 and 1727.

Finally, I apologise to the reader for the fact that the word 'experiment' occurs so frequently in these pages. I have avoided it where possible, but obviously it insisted upon being used in most cases. I must ask for indulgence. Any who finds such indulgence difficult to grant may permit me to offer in mitigation the report of Dr William Aglionby (F.R.S. 1667) in the spring of 1684 upon a German medical work:[9] he 'declared, that there was a great deal of reading in it, but little experiment.' At least I have avoided that error.

~ 2 ~

Aims and ideals

The primary aim of the Royal Society has never been in doubt, for it was recorded in the minutes of the first, preliminary meeting on 28 November 1660. Then those gathered in the room of Lawrence Rooke (d.1662), Gresham Professor of Astronomy, spoke of 'a designe of founding a Colledge for the promoting of Physico-Mathematicall Experimentall Learning', which it seemed might best be done by having 'a more regular way of debating things, and according to the manner of other countries' in order to 'the promoting of experimentall philosophy'.[1] The repeated emphasis upon experiment is the more worthy of notice because it was entirely original. By no means were all the foreign independent academies experimental in concept, while the Accademia del Cimento, which was, had been organised by, and worked directly under, private patronage and control. In contrast, the new English society was to be both experimental and independent, an organised continuation of private meetings that had taken place earlier, first in London in 1645 and later, then also in Oxford, now to be reformed with the addition of many who had returned with the King from exile on the Continent. The Royal Society was to be exceptional both in its attitude and in its procedures, while its emphasis upon experiment long remained its hallmark. As one of its first Secretaries, Henry Oldenburg, defined it in 1664, it was[2]

> a Corporation of a number of Ingenious and knowing persons, by ye Name of ye Royall Society of London for improving Naturall knowledge, whose dessein it is, by Observations and Experiments to advance ye Contemplations of Nature to Use and Practice.

This emphatic emphasis upon observation and experiment, and on their ultimate utility, was to be repeated by Oldenburg, as by most members of the Society, over and over again, and when Oldenburg died in 1677 it was to be repeated by successive Secretaries. At first sight, this seems merely an echo of the doctrine of Francis Bacon, and indeed many of the early Royal Society saw themselves as dedicated Baconians. Bacon figures centrally in the frontispiece to the semi-official *History of the Royal Society* (1667) by Thomas Sprat, and certainly Bacon's writings had been an inspiration of many of the founders (perhaps especially to Boyle). The Society interested itself in 'histories of trades' (accounts of crafts, agriculture, the mechanic arts generally) sporadically throughout the century, less, however, for practical reasons – to improve the methods used – than in the desire to accumulate facts and to learn more about the natural world, all part of the universal natural history which, so Oldenburg always told new correspondents, was one of the prime desiderata of the Society. Many of his fellow members would have agreed with Oldenburg, but it is very possible that by no means all would have put it, as Oldenburg did, as the principal aim of the Society. Yet most thought that it was important and all would have concurred that experiment and observation were needed for its practice.

That the declared aim of 'the improving [or promoting] of natural knowledge', which has been part of the Royal Society's full title since the Second Charter of 1663, could best be achieved by experiment and observation, none of the members ever doubted, even if they sometimes differed as to the manner in which these tools could best be employed. They by no means intended to exclude theoretical or mathematical science, and many were devoted to pure mathematics as well. Pure mathematics of course was hardly suited for discussion at meetings of men of diverse interests, many of whom were not learned but rather were virtuosi, men with an interest in the world of nature who enjoyed discovering what others were doing in the investigation of nature but who were not mentally or temperamentally equipped to investigate it themselves, except possibly in its simplest form, namely by collecting rarities or observing what went on around them. Mathematics, even those forms used in studying nature, was best put on paper. Theoretical science, if not too mathematical, it was possible to discuss at least to some extent, especially if it could be presented in conjunction with demonstrative experiment, and this was the easier since it was felt important by most Fellows to ensure that any theoretical principles were at least consonant with experiment, while many were anxious that their principles should be seen as clearly derived from experiment. Hypotheses, which most Fellows of the early Royal Society understood, as Newton was to do, as being theoretical principles not grounded in experiment, were best avoided as uncertain, although very many firmly believed that their hypotheses *were* confirmed by experiment. In any case, discussions of hypotheses, like discussions of politics and religion, were ever likely to lead to disputes and wranglings, upsetting the properly quiet atmosphere of learned debate, while experiment and observation could usually be spoken of without passion and dispute. In the early years of the restoration of the monarchy it was intended that the meetings should allow men of all political and religious opinions to gather together peaceably to investigate the natural world, turning their backs, at least weekly for an hour or two, on the social and political world. So royalist and parliamentarian, Catholic, Anglican and Puritan, all met together keeping their religious differences hidden (as the law demanded), while the social classes represented ranged from high nobility through the professions to well-to-do tradesmen – few at either end of the spectrum, but throughout the first sixty-five years at least of the Society's existence (the span of the present study) such a spectrum did exist.

By no means all of the early members of the Royal Society were, however, slavish devotees of the Baconian method, in spite of what has sometimes been written and was at times sometimes assumed for polemical purposes.[3] To most, Bacon was a source of inspiration to encourage an empirical approach to nature. This empirical approach was seen by some as demanding such a close attention to experiment and observation in the investigation of the natural world as, in extreme cases, virtually to exclude any consideration of theory. Others, perhaps most (and these included Robert Boyle and Isaac Newton), saw it as demanding an empirico-experimental basis for all theory and the consequent rejection of all hypotheses not clearly and rigidly based upon empirical evidence, as well as, sometimes, even using experiment to 'prove' theory. Others still, and these included Robert Hooke and such mathematically orientated figures as John Wallis and Christopher Wren, saw experiment as a most important basis for theory and investigation, but by no means the only one, and these allowed room for logically or mathematically derived theories. (It has been suggested that Hooke tended to stress theory *or* experiment, but not both at the same time.)[4] Parenthetically, a complication for the historian in determining the precise balance of theory and experiment in the work of seventeenth-century natural philosophers is that they often failed to perceive their own preference for such theory, providing that it was apparently supported by considerable experiment.

Hence it is that, to a certain extent, one must distinguish between aim and achievement, between ideal and reality. The aims are clear and were to be proclaimed over and over again. Indeed, as so often happens, they were proclaimed most loudly, as it seems, when it was felt that reality failed to reach the ideal. This led to review with the aim of reform and there exist a number of proposals for reform from different periods of the Society's early history. In some cases it is difficult to discern whether the writer clearly understood the difference between the aims and the reality. Thus Sprat, elected a Fellow in 1663 very possibly with the intention that he should write an account (a history in Baconian terms) of the Royal Society, as he almost immediately undertook to do under John Wilkins's direction, knew of the Society's work in the preceding years only through the eyes of Wilkins, while his examples of achievements drawn from subsequent years were selected with a Baconian slant, being especially examples of histories of trades or of inventions. The purpose of the Society Sprat defined[5] as

> to make faithful Records, of all the Works of Nature or Art, which can come within their reach, that so the present Age, and posterity, may be able to put a mark on the Errors, which have been strengthened by long prescription: to restore the Truths, that have lain neglected: to push on these, which are already known, to more various uses: and to make the way more passable, to what remains unreveal'd.

And as to method,

> I shall lay it down, as their Fundamental Law, that whenever they could possibly get to handle the subject, the Experiment was still perform'd by some of the Members themselves,

adding,

> it has been their usual course, when they have themselves appointed the Trial, to propose one week, some particular Experiments, to be prosecuted the next, and to debate beforehand, concerning all things that might conduce to the better carrying them on.

Sprat was not an active participant in meetings, as far as can be learned, and certainly he appears to write from hearsay, for both his tone and his matter evince that he is writing of the ideal. In actuality, as will appear, the Society's meetings were never so well organised as he suggests was the case, and of experiments proposed at one meeting only a small proportion were ever performed at the next or at all. Nor, if the minutes are to be trusted, did the meeting discuss the experiments beforehand, but rather it was left to the Curators (those who performed the experiment in question) to conduct such experiments as they saw fit. It is true that in the first years of the Society's existence many experiments were proposed at each meeting and, usually, Curators appointed for them. These Curators were expected to provide the necessary equipment and to instruct the Operator in his rôle as assistant or, sometimes, demonstrator. Later there were permanent Curators, the longest serving being Hooke, appointed in November 1662 and acting intermittently thereafter until his death in 1703.[6] Normally, individual members performed experiments at meetings as and when they chose, although quite often at the suggestion of others. And many experiments were discussed without being shown, accounts being provided by either Fellows or correspondents (cf. Figures 2, 3 and 4). Clearly Sprat's view of what occurred reflects the intention of the founders, especially of Wilkins, and what did more or less frequently happen in the first years when the Society was coming into being; by the time Sprat's *History* was published (1667) the more formal organisation of the Society had come into being.

An aim that never came anywhere near fulfilment was the ambition of establishing a 'College', that is of having a building of its own for the Society's use which should contain all the equipment necessary for carrying out work in all the branches of natural philosophy, from astronomy to zoology. As is well known, in pre-Royal Society days there had been many plans for 'colleges' devoted to many sorts of learning; of these the most relevant is that by John Evelyn promulgated in 1659, when he described in a letter to Boyle a strange vision of a truly monastic organisation for natural philosophers.[7] Many, especially the more ardent Baconians, fancied some such establishment in later years, but most, more rationally, hoped only for a home of its own for the Royal Society, forced throughout the seventeenth century to accept hospitality either from Gresham College or, more briefly, from the Howards at Arundel House. Occasionally it looked as if the Royal Society might be able to acquire its own establishment: in 1666 Oldenburg was to write optimistically to Boyle,[8]

> We are now undertaking several good things as ye Collecting a Repository, ye setting up a Chymicall Laboratory, a Mechanicall operatory, an Astronomicall Observatory, and an Optick Chamber.

(Of all this only the Repository, a collection of animal, vegetable and mineral objects, ever come into existence.) Similarly, Sprat was to suggest that besides Chelsea College, recently given to the Society by the King (but which it was glad to return to him a few years later, never having been able to secure satisfactory possession of it), the Society would soon build a college for meetings, laboratories, the repository, a library and lodgings for the Curators.[9] In fact, in the autumn of 1667 Wilkins initiated an attempt to acquire such a college by organising a committee of the Society to see to it; this got so far as to produce a form for subscription and to persuade no fewer than twenty-nine Fellows, not all by any means wealthy, to promise contributions, while Wren provided a design.[10] A few months later the scheme was dropped, never to be revived except in the modest form of a search for new premises at the end of the century when Gresham College finally refused to house the Society adequately with its demand for a meeting room, a library and a repository.

In any case, the provision of its own accommodation was not really central either to the workings of the Society or to its aims. What really mattered was activity by individual Fellows, and the means for best securing this was debated on many occasions. Several Fellows were to draw up and write out schemes for the reform of the Society to cure what they took to be its failings in respect to its perceived aims. One of the earliest examples is contained in a manuscript still in the Royal Society in the handwriting of William Neile (1637–70), F.R.S. 1662, son of Sir Paul Neile, one of the founders; William Neile had been a pupil of Wilkins and, as a promising young mathematician, a protégé of John Wallis, who in 1659 had published some of his original work in *De Cycloide*. In 1667 he had devised a theory of the tides, on which Wallis was working at about this time, and when theories of motion were discussed by the Society in 1668 and 1669, Neile's was one of those read at a meeting and registered, although unlike the theories of Wren and Wallis it was not published in the *Philosophical Transactions*. Neile was thus mathematically inclined, which makes his proposal all the more striking, since he there laid such stress upon experiment. His 'Proposalls humbly offered to better consideration' is undated, but presumably belongs to the later 1660s.[11] It is a carefully worked out statement with very definite conclusions, which begins firmly, 'The businesse of the Society is to make experiments.' Having said this, Neile continued by considering how these experiments should be planned and carried out. But he was not in favour of a simple Baconianism by any means, for once experiments had been performed he thought that they should 'be refered to a Committee to bring in a possible cause or causes,' because 'one of the cheife ends of the Society is to advance knowledge' and this can only be done by 'the indagation of the causes of these experiments'. Those who perform such experiments were, he thought, those best suited to find out their causes; and 'The bringing in of causes will very much advantage the carrying on the work of experiments.' Neile's belief in the value of committees was one held by most members of the Society, and from the very beginning of the Society's existence there had been committees set up on many subjects, although few lasted for very long. Debate in committee was easier and more effective than debate in open meeting, and Neile's view of their utility was that commonly held. Committees never ceased to be important in the organisation of the Royal Society and continued to exist throughout subsequent centuries. Nothing further, however, was heard of Neile's proposals: they were preserved in the Society's archives, but they do not seem to have been discussed either at ordinary or Council meetings, perhaps because of his death in 1670.

Some years later, in 1674, there were extensive plans for reform precisely because of the decline in experimental activity at meetings, and as will be seen below (Chapter 5) these had some effect in altering the management and content of meetings. In June of that year, when the Society was seen to be languishing with poor attendance at meetings, the President (William, Lord Brouncker) when adjourning the meetings until the autumn proposed[12]

> that in the mean time the Council might sometimes meet, and consider a better way than hitherto had been used, to provide good entertainment for the said meetings, by establishing lectures grounded upon and tending to experiment.

When the Council met in the autumn there was much discussion of the problem and (29 September) William Petty proposed that individual Fellows be appointed to entertain the meetings with experimental discourses, and to this end he further proposed that steps be taken to secure the money necessary to pay for experiments. (That experiments demanded equipment which the performer might not possess and that such equipment was expensive, even too expensive for many of the members to provide, was a novel perception, not often discussed then or now.) Petty was desired to draw up a plan to implement his proposals; he and Jonathan Goddard in fact collaborated to produce a joint 'declaration', the main emphasis of which was on means to secure payment of arrears of subscriptions, but whose chief effect was to turn the Society's 'entertainment' at meetings at least partly to lectures or discourses based upon experiment and to accept a diminution of experimental demonstration actually performed at meetings. This to a certain extent shifted the emphasis from *investigative* experiment to *demonstrative* experiment, weaving empiricism and theory closer together, as in many of the formal, published writings of the Fellows; and indeed many of the consequent discourses were to be published subsequent to their presentation to the Society.

Beside the drafts of Petty and Goddard, there exists an anonymous letter dated 19 October 1674, addressed 'Most honoured Sirs' (evidently to the Council) and signed 'Your most cordial Friend & most humble servant, A.B.' (The author was possibly Walter Needham).[13] A.B. (a common seventeenth-century alternative to 'Anon.') was insistent that 'Observation and Experiment' were the cornerstones of the Society's work. To this end he wanted the structure of the Society strengthened and, recognising that people work best when rewarded, wanted payments to be made to all officers as well as to lecturers (an interesting tribute to the possibly changing financial situation of the most active natural philosophers). He also wanted the King to be brought to take a greater interest in his Society, an aspiration never realised. Yet he deprecated the expulsion of members for non-payment of subscriptions (as happened periodically, notably in 1666, 1675, 1682, 1685, 1699), preferring that they should be ignored and penalised simply by not being informed of meetings. This paper was not discussed by the Council, presumably because what Hooke called Petty's New Model was preferred.

Certainly Petty's scheme was more realistic and, in so far as it concentrated upon trying to make members pledge payment of their subscriptions in advance, was more practical in financial terms, while its plan for discourses was both practical and useful. This plan for discourses, in fact, was the most effective aspect of all the proposals for reform: it was novel (although of course long letters, in effect discourses – like Newton's first paper on light and colours, or the 1668 papers on motion – had previously been read at meetings) and workable. The year 1674 saw the introduction of formal experimental discourses no longer disguised as letters and usually read by the authors; these were initially pledged in advance and the performers reimbursed for expenses, those who pledged discourses and failed to produce them being fined, thus helping to finance those who did duly perform. Evidently rumours of the reforms were spread in the coffee houses and elsewhere and interest ran high, for the first autumn meeting of the Society that year on 12 November had an exceptionally large attendance, 'Neer 40' being present according to Hooke. The plan was promptly put into effect and lasted as intended for about a year, with well organised lectures and discourses becoming a regular feature of the meetings. Even after the formal plan lapsed, the presentation of papers became commonplace and the custom continued, gradually over the next century replacing informal presentation and discussion.

In 1675 the smooth running of the Society was to be disrupted by bad feeling between Oldenburg, its conscientious Secretary, and Hooke, its hard-working Curator of Experiments, in a dispute that lasted even beyond Oldenburg's death in September 1677. The affair was rendered more disruptive because William, Lord Brouncker, as President, supported Oldenburg and carried the Council with him, to Hooke's frequently expressed disgust. (Briefly, the quarrel centred around the priority dispute between Hooke and Christiaan Huygens over the application of spring balances to watch movements, Huygens having designed and had built in practicable form what Hooke had suggested but not tried to construct many years before, and which he now accused Oldenburg of having 'betrayed' in correspondence with Huygens, which was not the case. The accusation was unjust, but Oldenburg and Brouncker were outspokenly, possibly prejudiced, upholders of Huygens.)[14] Hooke was so angry that in December of 1675 he began to think of founding a rival organisation, as he was to continue to do for the next year. He began with what he called a 'new clubb', which he seems to have hoped to construct around Wren.[15] For the last day of December 1675 his *Diary* records a long discussion with Wren and 'The beginning of a clubb for N[atural] P[hilosophy]'.[16] The 'New Philosophicall Clubb' seems to have met for the first time on Saturday, New Years Day, 1676; it was to be a secret organisation, its existence not to be revealed nor anything discussed at its meetings to be spoken of outside the membership. There were regular weekly meetings for the next six weeks and then, as is the way of such informal organisations, the meetings became less frequent and lapsed, reviving only briefly in the later summer. But by July 1676 Hooke had had a new notion, for he then recorded,[17]

> Contrived new Decimall Society of Boyle, More, Wren, Hoskins, Croon, King, D. Cox, Grew, Smethwick, Wild, Haak, for chemistry, anatomy, Astronomy and opticks, mathematics and mechanicks.

It is not clear whether this was to be yet another discussion club, or the new society on which he resolved a few days later (but at that time he also, most improbably, thought of emigrating to France), nor what the 'new club' he 'contrived' at Garaways Coffee House on 13 July has to do with any of the earlier clubs and societies he had thought of. He does record speaking with several of those named about his ideas: Lodowick (not named in the first list) 'liked it', Boyle apparently rejected it, Cox 'imbracd it'; but although Hooke records that he 'discoursd much about club' to 'Grew, Hoskins, Hill, Wren &c.' he does not indicate their opinion of it, and soon it disappeared from the *Diary*.

It is against the background of all this plotting and planning that one might view Hooke's well-known scheme for the Royal Society's reform, which has been variously dated.[18] His paper could have been written in the summer of 1674, when reform was being widely debated; it could as easily have been written after his quarrel with Oldenburg and the latter's death, when Hooke was active in the real re-organisation of the Royal Society, or at any time in between. Hooke's scheme, of which there exist several versions, tends to emphasise the advantages of cooperation almost on the pattern of Bacon's Salomon's House on the New Atlantis and the allied importance of persuading all members to work diligently to the common good. It stresses the importance of secrecy, as Hooke so often did, but at the same time emphasised paradoxically the importance of having correspondents at home and abroad. It also calls for the appointment of several Curators, each of whom should regularly bring in 'accounts & experiments' and produce regularly, once a quarter, a discourse with appropriately illustrative experiments, every member of the Society also to produce such discourses. This latter requirement would have meant restricting admission to the Fellowship to those who seemed, after careful scrutiny, fit to perform such tasks. This notion of requirements for admission to the Fellowship is one of the few serious proposals in the seventeenth or eighteenth centuries to exclude inactive members; normally they were even encouraged provided they paid their subscriptions to help finance the Society's proper activities.

There is no explicit reference anywhere in the Society's records to any of Hooke's proposals. And indeed they are not very practical for any but a small group, while his various clubs were necessarily restricted in numbers as small discussion groups, not institutions for the development, promotion and display of natural philosophy. Nor was Hooke himself in his later years so eager as he had been earlier to be held to the regular devising and performance of experiment in public, although he was always willing to discuss well-known experiments made by himself or others in the past, or potential experiments to be made in the future. Occasionally he could be stirred by matters raised at meetings to suggest or even produce one or two experiments to illustrate his ideas, but more and more his enthusiasm was directed towards discussion rather than practice. Once again, aim and reality diverged, even though Hooke continued to be interested in possible reform of the Society towards greater experimental activity.

Very possibly, Hooke, like other members of the Society, was influenced by the news of projected reform of the French Académie Royale des Sciences, which reached them by January 1699, as the reforms began. Certainly two months later the Council appointed a committee to consider the new statutes of their French colleagues 'and to make an Extract of what they think necessary to be Established for the Society.'[19] According to the minutes of the meeting of 15 March 1698/9 the fifteen Fellows present agreed 'to entertain the Society on certain days with some Experiments or Observations in Natural knowledge' and 'to deliver their observations, or leave them with one of the Secretaries.' As Sloane reported a week later, most of those pledging themselves to this task agreed to bear their own charges (presumably for experiments performed) although some wanted five pounds per annum, to which the Society agreed. In the event, only six of these Fellows provided their papers in the course of the year, namely Hooke, Houghton, Woodward, Havers, Pettiver and Sloane, while nine (Dowdy, Sister [? Silvestre], Cowper, Fatio, D. Cox, Moult, Robinson, Grew and Tyson) failed to do so; however, Moult did produce a paper in the next year. Success was clearly not as high as had been hoped, but not negligible. It was evidently, in view of those who did read papers, not related to the bad feeling which existed between several prominent members at this time, most notably between Sloane and Woodward, which generated a great deal of publicity between 1698 and 1700, and between the surgeon William Cowper and the Dutch physician to the King, Govard Bidloo, to name but two.

But these quarrels may help to explain some of the tone of the quite remarkable document entitled 'Proposals for the Advancement of ye R.Soc.', written after 1700, probably before 1703 but possibly as late as 1715.[20] It has been ascribed to Hooke, which would require a dating of

1701 or 1702, but is quite as likely from internal evidence (especially its lack of any particular emphasis upon experiment, always, at least in theory, dear to Hooke's heart) to be by someone else. It has an almost nineteenth-century ring in the scorn expressed for those who come seldom and 'as to a Play to amuse themselves for an hour or so,' who never think of 'Promoting the Ends of the Institution' yet, because they pay their subscriptions, 'take themselves to be ... good Members', as well as in its castigation of 'several Young Physitians' who have 'desired and obtained a Place', who may for some years have 'bin beneficial and promoted the Enquiries after Natural Philosophy [and] performed several excellent Matters' yet,

> as their End was only to get a Name for Extraordinary Men and procure a being taken Notice off and a large acquaintance for the increase of their particular Practice and advantage they give over participation.

Explicit, although even then qualified, praise if given only to the mathematicians (again unlike Hooke)

> from these the greatest helps have bin afforded and are to be expected – their particular studys giving opportunitys of new discoverys.

(I say qualified praise only, because they are listed immediately after 'Gentlemen who enter themselves for diversion only.') The stringent criticism of inactive members is sharply contrasted with what the author knew about the composition of the Paris Académie Royale des Sciences:

> The Parisian [Academy] consists of select Members all Pensioners paid for when they do their Tasks set and due performance Expected with at least a Philosophical scandal on the lazy and negligent.

The author of the proposals is still clearly proud of his own Society, claiming that it 'has the honour of being the first and giving at least a Rice [rise] to all the others.' He sees the necessity of pursuing a line different from that taken by the French, even though he is not very clear what this might be, for the proposal tails off unfinished with suggestions for the appointment of committees

> from Each of wch it was expected something should be produced and performed to be offered to the consideration of the body of ye society and their publique meetings.

It is remarkable that so much had the temper of the Society changed during the preceding decades that the word 'experiment' nowhere appears in this document. Perhaps the only possible explanation, other than change of purpose, might be the derision suffered by the Society during the years around 1700 for its practice of what seemed trivial and useless experiment and observation. (See Chapter 9.)

One more plan for the reform of the Royal Society exists from this period: that written by Newton. His 'Scheme for establishing the Royal Society' is unfortunately, like most of the plans for reform, undated; it has often been taken to have been written before his election as President but, especially as he showed little interest in the Society before his election as President, might well belong to 1713, when change was needed after Hauksbee, the then Curator of Experiments, died.[21] The 'Scheme' reveals more about the philosophy held by Newton himself than it does about the conduct of the Royal Society, being exactly in the same tone as his remarks about the proper practice of natural philosophy briefly expounded in his 1675 'Hypotheses of light and colours' and more fully in the General Scholium of the *Principia*, written in 1713. The 'Scheme' begins with a declaration of the true purpose and nature of natural philosophy, which

> consists in discovering the frame & operations of Nature reducing them (as far as may be) to general Rules or Laws, establishing those Rules by observations & experiments, & thence deducing the causes & effects of things.

The proposals then go on to suggest that certain of the Fellows 'well skilled in' various branches of natural philosophy 'be obliged by Pensions & Forfeitures ... to attend the meetings of the R.S.' so that they may be available to judge 'Books, Letters & Things presented to the meetings', where, it may be supposed, discussion will be encouraged. But the scheme ends a little inconclusively, with Newton declaring simply that

> The reward will be an incouragement to Inventors & it will be an advantage to the R.S. to have such men at their meetings, & tend to make their meetings numerous & usefull & their body famous and lasting.

Evidently, modern scholars who find the proceedings of the Royal Society's meetings in the late seventeenth and early eighteenth centuries lacking in solid natural philosophy and as having strayed from the design and practice of the founders are far from being anachronistic, for at least some of the members at the time saw the Society in just this light. Whether this last plan for reform was written before Newton became President (but how much did he then know about what went on at meetings, having attended so seldom?) or afterwards is not very important; what is significant is that he found the discussion which took place at meetings to be of a low level and he wished to raise it by making sure that there were always present some Fellows who could both judge the ideas of others and contribute their own ideas and experience. (In later years, Newton was to preside over many discussions of medical case histories, which may or may not have interested him, but to which he certainly sometimes contributed comments.) Rather curiously, since Newton's Presidency has been seen as, above all, a time of experimental performance at meetings,[22] there is in his scheme no mention of the performance of experiment nor of any need for the appointment of Curators of Experiment. The Newton of this paper wants expert judges but does not apparently envisage the need for Fellows to perform experiments at meetings. Once again, aim and reality diverge – for certainly until the last years of Newton's term of office, the performance of at least some experiment was characteristic of a preponderance of the meetings.

As it turned out, the Royal Society narrowly escaped becoming either a debating club or a copy of the formal Continental academies. In fact, it was never tied to state funds, nor did it separate into specialised sections. Rather it continued to be self-funded, even though poorly, and to expect its members to take an informed interest in the whole spectrum of natural philosophy. Hence the 'entertainment' at meetings might and did range from astronomy through natural history to medical practice and accounts of antiquities. But although experimental natural philosophy in all its branches remained the aim, the rôle played by experiment, as will appear, shifted greatly during the seventeenth and early eighteenth centuries. For from the showing of experiment, so important in the 1660s, the emphasis gradually shifted to the account and discussion of experiment and observation. This, as sociologists have noted,[23] raises the question of credibility, that is, the criteria by which the account of an experiment could be accepted by those who had not performed it. One can argue that natural philosophers had come to be more sensitive than they had been to the standing and skill of their contemporaries as regards their practice of natural philosophy. But it can also be argued that experiment had become so normal a part of natural philosophy, and its practitioners so accustomed to both perform and describe it, that a reader or auditor of an account could understand the experiment described so well that he did not see the need to view it directly.[24] If this is so, it argues for a greater sophistication on the part of the Fellows of the Royal Society in this period, and a maturing of the natural philosophy which they were endeavouring to improve.

~ 3 ~

The record of the minutes
1660 ~ 1674

It is helpful in considering the rôle of experiment to begin with the weekly activities of the Society as a whole, that is, by an analysis of the formal meetings.[1] In this way it is possible to discern what 'the promoting of experimentall philosophy' meant to the majority of Fellows (especially those who did not write books at great length), and it is even possible to quantify the proportion of experiment to discussion to some extent. And experiment was certainly at the heart of the Society's original activities as it had been for its originators, those then young men who, as John Wallis remembered in old age,[2] had since 1645 met with the same purpose and had even paid a subscription to cover the cost of experiment.

Before attempting to investigate the rôle of experiment at the weekly meetings in detail, it is well to consider the differing methods of presentation available for the promotion of experimental philosophy in the mid-seventeenth century. Sometimes a Fellow, who after the formalisation of the Society under the 1662 Charter was usually a Secretary, read letters describing experiments (although of course by no means all the letters read were concerned with experiment). Sometimes, and increasingly after 1674, Fellows read their own accounts of experiments performed elsewhere. Sometimes, and it is this which will be traced here, experiments were performed during the course of the meeting. In this latter case, it is commonly possible to know who performed the experiment in question, the Curator as it was called in the early days; he was usually assisted by the Operator, a paid employee. Only after the appointment of Robert Hooke as Curator in 1663 did the use of the word alter; Fellows continued to perform experiments at meetings, and both they and the official Curator were assisted by the Operator.

Something of what was involved is indicated in the minutes of the very first meeting on 28 November 1660, when 'Mr Wren [was] desired to prepare against the next meeting for the pendulum experiment.' To prepare for presenting an experiment meant assembling appropriate apparatus, planning a suitable display, trying the experiment to make sure that it would work, and perhaps providing an explanation and interpretation. Obviously appropriate equipment was necessary: one might possess it, or construct it, or have someone else construct it (sometimes the Operator), or buy it, or borrow it. Obviously, when showing an experiment to an audience, even a fairly small one, it is necessary to have a room where the experiment can be both effectively performed and visibly displayed. When, after 1645, meetings took place in London as described by Wallis, they required instruments as an adjunct to their discussions, and the performance of experiment supplemented their eager discussion. Indeed, in one account Wallis noted that there had been a weekly charge to cover the cost (which must have been either to purchase equipment or to employ an assistant). Because of the need for experiments, meetings were sometimes held 'at Dr Goddard's lodgings in Woodstreet, on account of his keeping an operator in the house for grinding lenses for telescopes'. Besides astronomical instruments and a skilled assistant, Goddard must have had a chemical laboratory, for he performed many chemical experiments at Royal Society meetings and described more.

At this first meeting, when Wren was asked to provide an experiment for the next one, it was agreed that in future the group would meet 'at Mr Rooke's Chamber in Gresham College, and in the vacation at that of Mr Balle in the Temple'. Rooke was now Gresham Professor of Geometry, having previously been Professor of Astronomy; the exchange of chairs, and the shift in meeting room, probably had the same cause, namely that the Professor of Geometry had better rooms than the Professors of either Astronomy or Physic (Goddard's chair since 1655).[3] Rooke without doubt possessed suitable instruments for use by the Society: as a notable astronomer at the least telescopes and lenses, while on 9 January 1660/1 the minutes record his being asked to supply tubes and quicksilver for the quicksilver experiment (i.e. for a Torricellian tube or barometer). Balle's chief interests were in astronomy and magnetism, especially the latter, and he, like Rooke and Goddard, possessed many instruments. It is not clear what happened after Rooke's death in 1662; he was succeeded by Isaac Barrow who was elected to the Society a couple of months after Rooke's death, so that perhaps he allowed use of his rooms as Rooke had done, although he did not often speak at meetings and may have been largely absent. When he resigned in 1664 the Society not unnaturally campaigned to have Robert Hooke, its recently appointed Curator of Experiments, elected to the chair. The choice of the College's Trustees fell on Arthur Dacres, a reasonably eminent physician but never F.R.S.; his only contribution to the Society's well-being was to resign in Hooke's favour after less than a year under pressure from the Society. And so once again the Gresham Professor of Geometry was able to arrange for accommodation, until 1666 (see below).

One other adjunct to the performance of experiments was almost immediately thought desirable: the appointment of an 'Operator' (with a salary of £4 per annum) to prepare apparatus and to assist any member whose experiment was being shown; where the Operator also helped to supply the equipment he was to be paid by the member concerned. Curiously, there is no indication that such an Operator was appointed for over a year. But there are references to an amanuensis, who was to copy out records and, after 25 February 1660/1, to attend meetings and 'set down such things as the Society shall think proper'. In fact by the next week he had begun to act as an Operator, being ordered to provide and prepare equipment for experiments when this could not be provided by Rooke or some other member. Some time in 1662 the situation seems to have changed, for although on 23 April 'The amanuensis was directed to take care to procure the long glass-tube, which Mr Colney promised to make', a week later it was the Operator who 'was ordered to provide against the next meeting two birds ... and a live mouse or two' for airpump experiments, while on 28 May and 4 June the Operator was to provide 'little fishes' to be kept in Dr Goddard's lodgings to make sure that they were looked after. For the next two or three months both Operator and Amanuensis seem indifferently to be asked to attend to both instruments and queries; it was not until the autumn that it is clear that the Operator was actually performing experiments at the Society's direction, although little is heard of him for some time after Hooke's appointment in November 1662 (see below).

Who were these employees? The minutes do not specify. After the granting of the second charter, when the Council first met on 13 May 1663, 'Mr Wicks was sworn as Clerk'. Michael Wicks had, in all probability, been the unnamed amanuensis for the previous two years; it is said that he had been one of Goddard's assistants,[4] which if true

President: Lord Brouncker

	(1)	Secretaries	(2)
1662–8	John Wilkins		Henry Oldenburg
1668–72	Thomas Henshaw		
1672–3	John Evelyn		
1673–5	Abraham Hill		
1675–7	Thomas Henshaw		

Curator of Experiments: Robert Hooke

Clerk: Michael Wicks

Operators
1663–76   Richard Shortgrave
1677   Henry Hunt

Editors of the *Philosophical Transactions*
1665–77   Henry Oldenburg
1677   Nehemiah Grew

Note: all dates are those of election, i.e. 30 November, unless otherwise stated; in the case of the editors of the *Philosophical Transactions* dates are those of volumes produced by them.

Figure 1.   Officers, 1662–77.

would explain his competence in experimental detail. Later in 1663 Richard Shortgrave, already known as an instrument maker, emerged as the Operator, a post he was to retain until his death in 1676, and it is possible that he was the unnamed Operator of earlier meetings. His work subsequently was often specifically assigned, and no doubt he performed many more experiments even than those for which he is named in the minutes, for many are noted as 'being performed' without any ascription; these he may have made and certainly he must have assisted many, perhaps most, of those named as performing experiments. He must also have kept the instruments belonging to the Society in order.

To return to what passed at meetings in 1661. Nearly three weeks after Wren had been asked to perform his pendulum experiment, it was resolved 'that Mr Wren bring in his account of ... [it], with his explanations of it, to be registered', which may or may not indicate that he had already performed it publicly at a meeting. At the same time,

> every member [was] likewise requested to bring in to a Committee, to be appointed for that purpose, such experiments, as he should think fit for the advancement of the general design of the Society.

Thus early it was perceived that experimental *demonstrations*, although interesting and indeed necessary for enlightenment and general interest, were ephemeral, and written accounts were required for the preservation and transmission of both the facts of the experiment and the circumstances surrounding it (which, as Boyle was at this time noting in essays on successful and unsuccessful experiments, was essential information) as also of its explicit purpose. A Register Book was soon inaugurated to preserve the accounts provided, which most Fellows were punctilious about preparing at this time, although less so later, and which served a double purpose as they could be used to establish priority. Experiment performed before an audience was instructive to both audience and performer, for discussion could improve its presentation and suggest application to theory, as also variations. There is little indication that reports of experiments by Fellows who had performed them were held to need demonstration to confirm reliability, although reports of experiments from non-Fellows at home and abroad often were held to need such confirmation.

There was yet another, lesser, purpose in experimentation. This was to entertain. All members enjoyed seeing experiments, and especially ingeniously devised or adroitly performed experiments, whether these were highly instructive or not. Additionally, it must be remembered that there were many members who were virtuosi rather than practising natural philosophers. And virtuosi, by definition, *enjoyed* natural philosophy. For them experiment provided far more interest and entertainment than the reading of formal papers. This was even more true of visitors, who were often led by seeing the experiments to a desire to become members. (And by paying their subscriptions, if they did, they helped to provide money to pay salaries and wages and the costs of experiment.) Thus Pepys had visited the group early in its existence (January 1660/1) and heard many accounts of its experimental activities before being elected a Fellow in February 1664/5; he always enjoyed attending meetings, especially when good experiments were made, although, as he confessed to himself, he lacked the philosophy to understand the explanations of them. And when plans were laid for the entertainment of distinguished visitors – from Charles II, who never came, to the Duchess of Newcastle and the Tuscan Ambassador who did – it was experiment which was intended to provide the entertainment.

By 2 January 1660/1, after only four meetings, the proceedings were already assuming their future pattern, with plans for experiments dominating. So,

> Mr Boyle was requested to bring in his cylinder, and to shew at his best convenience his experiment of the air [already familiar to his Oxford colleagues]; as Dr Merret was to bring in the history of refining; Dr Goddard his experiments of colors; and Dr Petty the diagrams of what he had discoursed to the Society that day, and the history of building ships,

while Brouncker and Boyle were to provide a list of questions (directions for observation and enquiry) for travellers to the Peak of Teneriffe, then the highest readily available mountain. Here is an excellent example of the variety of empirical interests among the Fellows: physical and chemical experiment, accounts of industrial practice and of inventions, accounts of natural history. And here too is a good example of the variety of presentation possible: the showing of experiments, the presenting of accounts of experiments, illustration of 'discourse' (here probably informal), and the reading of papers.

On the whole, it was experimental demonstration which dominated the meetings in 1661. Exceptionally, on 16 January,

> The King sent two load-stones by Sir Robert Moray with a message, that he expected an account from the Society of some of the most considerable experiments upon them,

exceptionally because of the royal interest, but not otherwise, for any member might have, and later did, initiate experiments with loadstones. The result of the message was the appointment of a committee to perform experiments, discuss them and report on them, which was also normal practice. During this year some two dozen individual experiments were made at the meetings, while a dozen or more made elsewhere were carefully described orally. About a dozen experiments were 'ordered' by the assembly to be performed at future meetings, while over three dozen were ordered to be performed elsewhere. The choice of whether to perform experiments at meetings or elsewhere was at least in large part dictated by convenience: because it was more convenient for the Curator of the particular experiment to work in rooms of his own choice, or because a series of experiments was required, or because the experiments were not suitable for performance in the meeting room (as, for example, ballistic experiments, falling body experiments, or barometric experiments demanding changes in height, requiring hills or mines or tall buildings, or some anatomical experiments). It is unfortunately not possible strictly to correlate precisely experiments ordered and experiments performed. Particularly was this the case when the person who did perform relevant experiments was not named, but those who were named as performing experiments include Boyle, Brouncker, the amanuensis Michael Wicks, Croone, Goddard, Moray, Wilkins, Henshaw, Povey and Greatorex. (Which Mr Henshaw is not certain but probably here Nathaniel, an Original Fellow, d. 1673, who wrote on meteorology, nitre and gunpowder; Ralph Greatorex, an instrument maker who worked for Boyle before 1660, was never a Fellow, although he had been associated with the pre-Society London meetings. Those named as reporting on experiments made elsewhere were Boyle, Brouncker, Wicks, Moray, Goddard, Henshaw, Sir William Pearsall, Balle and Rooke, all, except Wicks, to be Original Fellows, that is, named as such in the 1663 Charter.)

Even in 1661 there were beginning to be reports of experiments, often sollicited, received from those not as yet part of the organised society. Thus in May, Croone, appointed Register, was asked to write to Dr Henry Power 'and to procure a correspondency between Sir William Pearsall, Mr Balle and Dr Power' on magnetic experiments, in which all three were known to be interested. This resulted, a few months later, in Power sending an account of experiments 'on mercury' (that is, made with the barometer or Torricellian tube) 'at the bottom and top of Halifax-hill'.[5] These experiments, performed by Power with his friends the Towneley family, all active virtuosi, were the first of Power's papers to be read at the Royal Society, to be published later in his *Experimental Philosophy* (1664). During the next year, Power was to send many magnetical experiments, as requested, together with experiments on weighing objects above and below ground – these latter also initiated at the Society's request (as in 1663 he responded to the request that he observe a lunar eclipse). In June 1663 he came to London and began (24 June) to attend meetings, and it was only then that he 'produced several microscopical observations made by himself'. As a result, 'Dr Wilkins, Dr Wren and Mr Hooke were appointed to join together for more observations of a like nature.' (It should be noted that there is no indication that Power's observations were in question; it is rather that they were so interesting and so well approved that more were desired.) The result was that Hooke, now Curator of Experiments (see below p. 31f.) showed over the next months most of the microscopical observations described in *Micrographia* (1665). But to return to 1661: there were a number of anatomical experiments, inevitably of great interest to a group in which there were so many practising physicians, many accounts of anatomical abnormalities, and some medical and surgical case histories.

The next year, 1662, began with considerable interest in providing ship-captains and travellers with 'inquiries' about foreign parts; these, which included queries about climate, natural history of plants, animals and minerals, technology, customs of the inhabitants, and so on, were intended to provide the empirical basis of the long-planned but never completed 'universal natural history' which the Society hoped to compile. But the main activity continued to be the performance of experiment before the Society and the reporting on experiments made elsewhere. Among those who provided experiments and experimental accounts were Boyle and Goddard (over a dozen each), Brouncker and Croone (half a dozen each), and, as noted, Power. A new name, that of Robert Hooke, now appears: he performed some ten experiments at the very end of the year. His appearance is related to the formalisation of the Society, which became the Royal Society officially with the granting in the summer by the King of the First Charter. This named a President (Brouncker), two Secretaries (Wilkins and Oldenburg), a Treasurer (Balle) and a Council.

It must already have been obvious that considerable experimental activity was going to be difficult to obtain, when it had so far been confined to a relatively small number of the members (now denominated Fellows), in fact about fifteen, and no doubt it was becoming something of a burden to these, most of them men busy in the world of affairs or of medicine. However this may have been, plans were now made for the establishment of the permanent post of Curator of Experiments, to be held by someone worthy to be a Fellow but who, like the Secretaries, should (in principle) receive a salary from the Society. In the words of the Journal Book for 5 November 1662,

> Sir Robert Moray proposed a person willing to be employed as a curator by the Society, and offering to furnish them every day, on which they met, with three or four considerable experiments, and expecting no recompense until the Society should get a stock enabling them to give it.

(It should be remembered that the Society's only funds came from the Fellows' weekly subscriptions and that many failed to pay them.) The minutes further note that it was unanimously agreed to 'Mr Hooke being named to be the person,' and he was formally proposed to the

Society as Curator a week later, when 'it was ordered, that Mr Boyle should have the thanks of the Society for dispensing with him for their use', Hooke having for some years been Boyle's own assistant. (As the Society claimed to have no money to pay him, it must be presumed that Boyle continued to do so at least for the next year; it is certain that Hooke continued to regard Boyle as his patron.)[6] Hooke was known to most of the members of the Society as Boyle's able assistant and also as the author of a little tract on capillary action (1660) which the Society had intended to debate at the meeting of 1 May 1661 (the minutes fail to record that they did so). So at the end of November 1662 it was officially ordered

> That Mr Hooke should come and sit amongst them, and both bring in every day of the meeting three or four experiments of his own, and take care of such others, as should be mentioned to him by the Society.

This may seem a heavy burden, but it is clear that only simple experiments, quick to perform and designed only to stimulate discussion or even merely to entertain, were initially envisaged; in fact in the remainder of the year Hooke showed three or four experiments and reported on seven others, activity with which the Society was well content. He was soon, as intended, elected into the Society: in May 1663 the Second Charter lists him as one of the Original Fellows, although his position and salary were regularised only on 27 July 1664.[7] He was indeed a perfect choice; his versatility, his love of experiment and his facility in devising suitable experiments for public presentation being all notable. Once his attention was turned to a subject, he could seemingly always quickly produce a noteworthy comment, whether theoretical or experimental; added to this his ingenuity as a deviser of instruments, his astronomical competence and his taste for new hypotheses and theories meant that he was able to contribute to almost any subject, whether physical, astronomical, physiological, chemical or anatomical. Further, he far preferred to talk about natural and experimental philosophy to writing books about it, so that the Society's meetings offered him the perfect setting. (True, he did later sometimes complain of the burden, but this was more because he thought himself unappreciated or denigrated than because he disliked the post.)

During 1663, in the first flush of enthusiasm, Hooke provided at least a quarter of the experiments discussed at meetings, performing over a dozen as well as reporting on many more and also showing the results of his microscopical investigations, which the Society constantly urged him to continue and prepare for publication, and it is impossible to judge in how many of the two dozen experiments whose 'Curator' is unnamed involved Hooke. It must be noted that the Operator was specifically charged with some experiments, whereas in 'ordering' experiments for the future at least a dozen Fellows besides Hooke were named as Curators. Among these, Goddard's name figures largely, and both Boyle and Brouncker were responsible for at least half a dozen experiments each. An interesting interlude occurred in the summer of 1663 when the Society tried to reproduce Huygens' discovery (first made by him in 1661/2) of the anomalous suspension of water in a Torricellian tube in an airpump. At the first trial (1 July) it failed, but on 19 August it succeeded excellently, after which Hooke was asked to write an account. Further, at the Society's request, Brouncker had the experiment repeated (by whom is not stated) and found in several trials that it succeeded, and the Society thereupon accepted the experiment as a reliable one. This is an excellent example of the operation of the Society's firm belief in its motto of *Nullius in verba*; even when experiments were reported by so reliable a Fellow as Christiaan Huygens (F.R.S. 1663) they were subjected to scrupulous repetition, confirmation and analysis in this early period.

In 1664 the Society met faithfully every week, save only for that in which Ash Wednesday fell. It was a bumper year for experiments, about seventy-five being performed at meetings and the same number 'brought in', that is described as performed elsewhere. Most of those performed at meetings are not ascribed to any one member in the minutes. There are exceptions: Hooke is named for five, possibly those he proposed himself, although he is recorded as reporting on fifteen performed elsewhere; Boyle is recorded as performing three and reporting on five; Croone as performing three, the Operator two and Charleton one; at the same time a dozen Fellows (including Charles II, who weighed himself before and after exercise) described verbally or in writing experiments they had performed elsewhere. Besides this, Oldenburg as Secretary frequently read reports from his increasingly vast official correspondence at home and abroad; these letters, filled with both theory and experiment on all manner of subjects usually gave rise to lively discussions which produced more reports of past experiment.

Never again did the Society meet so continuously, but customarily intermitted its meetings in the late summer and early autumn, and it was long before the meetings were so well attended as they had been in 1664. Although 1665 began well, plague broke out in England in May 1665 and was epidemic in London by June, which inevitably and sensibly sent many members out of town. On 20 June the Council ordered that 'by reason of the present contagion' the President should propose that 'it would be convenient to intermit their publick weekly meetings'. This he did the next day, there being one more meeting only (28 June). During the difficult period that followed, Brouncker and Oldenburg stayed in London, from whence the latter kept up his correspondence with learned men. The Court went to Oxford, and with it went Moray and other courtiers to find Boyle still living there, as well as the Savilian Professors of Geometry and Astronomy (Wallis and Wren, respectively), and the Sedleian Professor of Natural Philosophy, Willis, whose pupil Richard Lower was working with Boyle and Wren on physiological experiments soon to be of great interest. Hooke, who had been 'ordered to prosecute his Chariot-wheels, watches and glasses during the recess',[8] stayed in London until early June, whence he wrote to Boyle.[9] As he said, he intended, with Wilkins and Petty, to go to Nonsuch (near Epsom) 'to prosecute the business of motion', taking the Operator with them; he also proposed experiments on the resistance of fluid mediums and was willing to undertake anything proposed by Boyle. In fact they went to Durdans,[10] 'my lord Berkley's house near Epsom'; Wilkins and Petty soon left, but Hooke stayed until the beginning of February 1666 when, the epidemic subsiding, he returned to his rooms at Gresham College. By then most of the Council members normally resident in or near London had also returned.

When the Society resumed its meetings on 14 March 1665/6, various books published in the preceding months were presented, after which Brouncker inquired of those present as to their 'employments' during the recess. Wilkins and Hooke spoke of the trials they had made to improve 'chariots', and Wren and Hooke were ordered to try that actually built. Hooke also spoke of his experiments on weighing bodies 'in a very deep well and above ground', when no differences had been found.[11] Moray gave a long account of his experiments on Welsh lead ore, but confessed that his long-promised history of masonry was still incomplete. Daniel Cox had examined the crystalline structure of salts, which aroused so much interest that

> He was urged to go on vigorously in so noble a subject; and to desire in it the conjunction of Mr Boyle, Sir Robert Moray, Mr Henshaw, Dr Goddard and Mr Hooke.

(Hooke and Boyle had both been interested before in this subject, Henshaw had read a paper on saltpetre in 1661, and Goddard's chemical interests were life-long.)

Most novel of all was the subject of injections into the veins of animals, a subject not previously brought up at meetings but one of European-wide interest. Dr Timothy Clarke was asked for an account of his trials; he had read a paper on the subject on 16 September 1663 using a technique in fact devised by Wren (as Boyle had reported in print in that year in Essay 2 of his *Considerations Touching the Usefulnesse of Experimental Natural Philosophy*) and tried by Willis and Lower in 1661–2.[12] The Society's present interest arose from reports of experiments conducted abroad, in both Germany and Italy. It is not clear why Clarke claimed a share in developing this technique. Now he merely

> gave answer, that he had not neglected it, and intended to finish it, as soon as possibly he could, for the press,

but in fact the only thing he wrote was a long Latin letter for Oldenburg to publish in his *Philosophical Transactions* for 1668, a history of English endeavours, which claimed English priority.[13]

Even to this first Society meeting of 1666 it must have been apparent that little was to be expected from Clarke, and Moray was moved to speak of the more ambitious experiments performed at Oxford under the aegis of Boyle (but in fact by Richard Lower) of 'transfusing of blood of one animal into another', which he denominated 'a considerable experiment if it could be practised'. Clarke denigrated this achievement, saying that he himself had tried it two years earlier and, not surprisingly, found it too difficult, to which Moray countered that 'Mr Boyle was of opinion, that the difficulties of this experiment might be mastered'. Indeed, Lower was, at this very time, successfully performing just such an experiment in the presence of Wallis, Dr Thomas Millington and other Oxford doctors. The Society was keenly interested: at the meeting of 18 April Boyle was asked for more direct information but modestly saw to it that Lower himself should send a full description, as he did in a letter dated 6 July, read to the Society in the early autumn.[14] Later, in *De Corde*, Lower was to describe the way in which he arrived at his final, successful technique, avoiding clotting by using live subjects and following nature, as he put it, by transfusing the blood from an artery of the donor to a vein of the recipient. In his letter he described how to lay bare and ligature the cervical artery and the jugular vein, and insert quills to act as tubes to convey the blood (two in the recipient animal, since it was assumed that blood must be withdrawn in proportion as blood was inserted). Dogs were used; the donor was bled until it died, whereupon the recipient was sewn up and when set free showed itself very lively. Lower noted several possible improvements: the use of specially prepared silver tubes in place of quills, which would make it easier to ensure that the connection between blood vessel and tube was really tight, and the use of an artery taken from a horse or ox to make a flexible connection between the animals.

So novel a technique naturally aroused the greatest possible interest within the Society and, ultimately, abroad. As a result of Lower's description, Daniel Cox, Thomas Coxe and Hooke were appointed Curators of this experiment, which they were asked to try in private before repeating it, if they were successful, before the Society. The Committee was, as so often happened, a little slow to act, and they naturally did not find the experiment easy to repeat, but they did in the end succeed, so as to be able to repeat it at a meeting. The minutes of 14 November 1666 record that

> the experiment of transfusing the blood of one dog into another was made before the Society by Mr King and Mr Thomas Coxe.

(It is not recorded when Edmond King, an expert anatomist and F.R.S. July 1666, was added to the team.) Now that the Society had been assured of the potentiality of the experiment by 'ocular demonstration' (even Boyle's vouching for the truth of Lower's claim for success was not enough in so difficult and odd an experiment), it was felt proper to let it be made public. So Oldenburg was authorised to make a last minute insertion into no. 19 of the *Philosophical Transactions* (dated 19 November), proudly declaring

> This Experiment, hitherto look'd upon to be of an almost unsurmountable difficulty, hath been of late very successfully perform'd not only at Oxford, by the directions of the expert Anatomist Dr Lower, but also in London, by order of the R. Society, at their publick meeting in Gresham Colledge: the Description of the particulars whereof, and the Method of Operation, is referred to the next Opportunity

which was in the next number (no. 20, 17 December 1666, pp. 353–7). King's account of a later experiment of transfusion between a calf and a sheep (described to the Society on 4 April 1667) was to be published in the *Philosophical Transactions* only a month later. Meanwhile, the Society on 12 December 1666 having ordered the trial of transfusing the blood of a sheep into a dog, this was reported on by Thomas Coxe on 21 March 1666/7. There had been a hiatus in such experiments during the cold winter weather; the revival of interest in the spring came partly from the warmer weather, partly from the news of similar experiments made abroad, notably in Paris: an abstract of an account by Jean Denis who had performed the operation, the account published in the *Journal des Sçavans*, was read to the Society on 21 March 1666/7, although news of his achievements had reached the English earlier.

It must be emphasised that at this point there was no question of rivalry nor of a dispute over priority. Indeed Oldenburg's intention in publishing Lower's account had been 'to invite others to ye like Experiments'.[15] Priority questions, however, arose as a result of a long letter by Denis, also published in the *Journal des Sçavans*, in which he not only described transfusions made upon two men, using animal blood, but claimed priority for the idea of blood transfusion.[16] As far as concerns *human* blood transfusion, the French did have priority, as it was first tried in England on 23 November 1667 at Arundel House by Lower and King, both of whom had been practising animal transfusion over the preceding months. (This experiment was shown to many spectators, although not at a meeting of the Society.) The subject, one Arthur Coga, an indigent Oxford graduate, not only survived being given the blood of a sheep, but two months later he read a paper to the Society describing the effects which he had experienced. The details of the transfusion naturally differed from those performed on dogs, and Coga received the blood in his arm. Clearly both Coga and the experimenters were very lucky and lucky also in not attempting a repetition, which would probably have been fatal. One of Denis's subjects did die after a second transfusion, whether, as seems likely, as a direct result of the experiment or, as Denis claimed, from poison administered by his wife. In any case human blood transfusion was then legally banned in Paris and the Royal Society wisely decided to discontinue the practice.

By this time, in fact, King's attention had already been turned to experiments upon respiration, experiments originally designed by Hooke, in which the thorax of a dog was opened and the animal kept alive by blowing into the lungs with a bellows, a horrid experiment which Hooke later disliked and refused to repeat, but a very instructive one;[17] Hooke and Lower had jointly performed it at the meeting on 10 October 1667. (This was one of the experiments which Lower cited in *De Corde* as evidence for the rôle of the air in turning blood red as it passed through the lungs.) Physiological experiments by Lower, as well as others, continued to interest the Society. Indeed the Fellows were so much impressed by Lower's achievements and dexterity in experiment that at the Council meeting on 5 November 1667 it was suggested that he be asked to become a Curator of anatomical experiments. But nothing came of this.

This long series of physiological experiments, some performed at meetings of the Society, some performed elsewhere and reported upon, shows what could be accomplished when initiators were prepared to take the trouble to work out the consequences of an original innovatory experiment. Here several Fellows, most notably Lower, pursued an experimental idea common to several of them, and pursued it for many months, carrying it to a logical if, as regards human blood transfusion, premature conclusion. Here was real cooperative research such as was more common in the Académie Royale des Sciences, just coming into being, than in the Royal Society.

In considering experimental activity in 1666 it must be borne in mind that, like 1665, it was a year of catastrophe and the Society's meetings were, of necessity, relatively few. As already noted, meetings had only resumed in mid-March. The Society then met regularly during the summer, only to be prevented from meeting for a fortnight at the beginning of September by the 'late dreadful fire', as the minutes call it, which began on 2 September. On 12 September, the Council met at Gresham College in the rooms of Dr Pope, Professor of Astronomy, after which the Society was forced to leave Gresham College entirely, although Pope retained his rooms. Charles Howard (F.R.S. 1662) offered rooms in Arundel House in the Strand,[18] less suitable (Arundel House was still a very old fashioned huddle of buildings around a courtyard) but possible. And there the Society was to remain until December 1673, when it was at last able to return to Gresham College. Although the accommodation must have been less suitable, Arundel House was more convenient as a meeting place for those not Gresham professors but courtiers or lawyers and for those like Boyle and Oldenburg who lived in the newly built up and newly fashionable area near St James's Palace. When the Society did return to Gresham College, it obtained spacious rooms with a library, a meeting room and a room for the repository.

In view of the number of experiments actually performed at meetings in these years, especially during 1666, and what, admittedly with hindsight, can be seen as the singular persistence of the medical Fellows in pursuing difficult physiological experiments, it cannot have been because experimental interest had lapsed but rather other causes that promoted a review by the Council of experimental procedure at meetings. It may have been the result in part of the move to Arundel House, where there was no great collection of instruments; it may have been the very success of the experiments of the previous months which had been developed over a period of time with several Fellows working together.[19] In any case, at the Council meeting of 4 December 1666 Moray suggested

> That the Council would take into consideration, how the experiments at the public meetings of the Society might be best carried on; whether by a continued series of experiments, taking in collateral ones, as they were offered, or by going on in that promiscuous way, which had hitherto obtained.

It is probably not significant that no more discussion is recorded, for such proposals had a way of lapsing, although they were perhaps pursued unofficially; more significantly, at the same meeting there was raised the query 'whether the experiments for propagating motion and the magnetic ones, should not be prosecuted'. Nothing was definitely planned for the future, but it was noted that Huygens and Balle 'had engaged themselves particularly the one in those on motion, the other in those on magnetism'. William Balle had often in the past been a Curator for magnetic experiments and had borrowed apparatus from the Society in order to pursue them, as he continued to do as late as 1680.[20] Huygens equally continued to pursue the problem of motion, as was to emerge in 1668 when the topic was actively discussed at Society meetings.

An important distraction from individual contributions to the Royal Society's programme of experiments arose from the need to rebuild London after the Great Fire of 1666. Wren's scientific interests had been overshadowed by his architectural interests at least since 1663, when he had been involved in a scheme to rebuild Old St Paul's, and now this and other projects came to occupy him. Not until the 1670s did his scientific career finally give way before his architectural interests; he remained active in the Society's administration for another ten years but his attendance at meetings became less and less regular after 1669 and he made no more original contributions. Similarly, Hooke also began to devote considerable energy to architectural matters: he naturally had a plan of his own for rebuilding London, was in fact appointed City

Surveyor as early as 1666, and between 1668 and 1674 was particularly busy designing and supervising both alone and with Wren. It is therefore no wonder that his activities for the Royal Society tended to slacken a little, nor that he much wanted a personal assistant (Henry Hunt was appointed only in 1673).

The first half of 1668, like 1667, had been dominated by physiological experiments, performed chiefly by Lower, King and Hooke. But there were many other topics 'prosecuted'. At the beginning of the year there was a great deal of interest in instruments: Towneley's dividing box, various horological devices invented by Hooke, his weather-glass for measuring the pressure of the air at sea, his new cyder-engine of which it is noted 'he was put in mind ... and ordered to get a model of it made',[21] Croone's 'wind gathering vessel',[22] and many more. The number of experiments mentioned as performed was much the same as the number recorded for 1667; far more of those actually performed at meetings have no name attached to them so that it is impossible to say who was responsible. In the case of those merely described (i.e. performed elsewhere), the names of King, Boyle and Hooke are those most frequently mentioned. The many letters from Fellows and well-wishers, both at home and abroad, which contributed greatly to the content of the meetings might contain experimental facts but more often were descriptive of observations, the most common being astronomical, or contributions to the 'universal natural history' which the Society actively sought. Although the performance of experiment before the Society did not greatly languish compared to the previous year, it was well below what was desired, few of those experiments 'ordered' being actually brought in. Hence presumably the Council's request (13 April) that the President should

> signify to the Society, that considering the want of experiments at their public meetings, the Council had thought proper to appoint a present of a medal of at least the value of twenty shillings to be made to every Fellow, not Curator by office, for every experiment, which the president or vice-president shall have approved of.

The President was to consult with the Master of the Mint about having the medals made, but no more was heard of the plan.[23] Ruefully, the minutes of the next meeting (16 April) record that

> The experiments appointed for the next meeting were the same, which should have been made at this, but were not,

these being 'weighing metalline bodies', the improvement of a cyder-engine, and the procuring of 'optic glasses' (made by both Christopher Cock and Hooke) 'for seeing in the dark'. The hydrostatic experiment was in the end not tried on 23 April because the balance needed adjustment and accurate weights were not available, and its performance was indefinitely postponed. This provides a good example of the problems facing those who wished to perform experiments in public and in ordinary rooms, especially those involving delicate apparatus. When (30 April) Boyle was asked for information about the experiments upon which he had been working 'during his late absence' (caused by ill-health) and, having briefly described them, was asked to 'communicate some of them to the Society' it was not his fault but the delinquency of the Operator which prevented their immediate trial. King was always ready to repeat the experiments which he described in accounts to the Society, in which Hooke often assisted him, perhaps because Hooke was more adept in public presentation. Both Boyle and Hooke were continuously active in describing and proposing suitable experiments. On balance, the record of experiments is quite high for the first seven months of 1668, until the Society's adjournment at the beginning of August.

When the Society reconvened in the autumn (22 October), Brouncker tried to persuade Hooke to tell them about his 'attempt to prove the motion of the earth from observations' (as he described the Culterian Lecture he was to publish in 1674), but Hooke, unwilling, diverted attention from this by proposing 'that the experiments of motion might be prosecuted, thereby to state at last the nature and laws of motion'. (Such experiments had been tried some years since by Brouncker, Hooke and others.) Brouncker demurred, declaring that he

> desired, that it might be considered, whether it were so proper or necessary to try this sort of experiment, since Monsr Huygens and Dr Christopher Wren had already taken great pains to examine that subject, and were thought to have also found a theory to explicate all the phaenomena of motion.

Although Hooke may have taken this as a snub, the other Fellows present seem to have agreed with Brouncker, for Oldenburg was ordered to write to both Huygens and Wren[24]

> to desire them, that if they did not yet intend to publish their speculations and trials of motion, they would communicate them to the Society for their consideration, and be assured, that they should be registered as their productions.

This is a clear statement of several rôles which the Society could play: it could encourage publication, or failing that it could encourage formulation of theories and discussion of their merits, always preferably with their experimental basis, and it could record possession of ideas and by implication establish priority. It could also suggest and/or perform experiments to confirm or explore such theories. Thus Hooke at the next meeting

> moved, that experiments might be made to see, whether all hard bodies, that rebound, do not so upon the account of having springy particles in them; and that it might be inquired into, whether there be any body springy upon any other score, than that it has air in it,

for he believed that an absolutely hard body would not of itself be elastic. Others pointed out that such an experiment could not be performed without trying an absolutely hard body, certainly not an easy thing to come by, and although Hooke proposed as an alternative that bodies of varying hardness might be tried, he soon gave up. Instead (5 November) he proposed 'the trying of experiments to determine the question concerning the communication of motion', promising to pursue such experiments by means of wooden balls suspended on strings. In the next two weeks he showed several experiments on inertia, on rebounding bodies and, in the New Year, 'made an experiment to prove, that the strength of a body moved is in duplicate proportion to its velocity'. Meanwhile, Wren had sent in his paper of the theory of collision, whereupon Hooke was asked to repeat the experiments which Wren described, some of which he did in the next month. By this time the Society had decided to survey the whole subject of motion, especially of impact motion. When (14 January) Croone 'brought in his hypothesis of motion', it was decided to ask interested Fellows to compare this with other theories available – ultimately those by Wren, Huygens, Wallis and William Neile, all of which were very similar.[25] (Indeed, Huygens 'acknowledged Dr Wren's laws of motion as altogether conformable to his'.) Early in 1669 the Society tried to secure further experiments from both Hooke and Croone, but although Hooke performed one or two more, both his and Croone's interest had been diverted to other things. As this suggests, the Society could fairly easily arouse interest in an experimental subject, but could not always sustain such interest long enough to reach any consensus or conclusion; once several serious accounts had been presented on a topic, the members of the Society tended to lose interest in it and to turn to something else. So when two years later (1671) the young Leibniz sent his treatise on motion, *Hypothesis physica nova*, it provoked no discussion at meetings, for both Wallis (to whom it was sent) and Hooke found little originality in it; they either regarded his ideas as well established or, as probably incorrect.[26]

To return to 1669: the number of experiments shown in the first part of this year was only a little lower than in 1668, Hooke indeed performing more than in the previous year. Nevertheless, there was some dissatisfaction with the experimental content of the meetings, perhaps partly because there were few reports of experiments performed elsewhere to supplement those displayed at meetings. At any rate, at the Council meeting of 1 February 1668/9 Oldenburg moved

> that the Council would think upon an effectual way of carrying on the business of experiments at the meetings of the Society,

suggesting the formation of one or two committees to 'direct' experiments. Six weeks later, the Council 'declared, that it was necessary to have another curator', and it was agreed that Hooke might hire an assistant, if he could find one to suit him. It seems very possible that Oldenburg was prompted by that paper, already referred to, entitled 'Proposalls humbly offered to better consideration' which begins, firmly, 'the businesse of the society is to make experiments,' goes on to suggest, among other things, Committees to examine all experiments and rules for procedure, although the paper also declares 'the experiments themselves are but a dry entertainment without the indagation of causes'.[27]

The failure to find either a second Curator or an assistant to Hooke must partly explain the sharp decline in the number of experiments either proposed or performed during 1670. Hooke performed only about a third of the number for 1669, and other Fellows were not diligent. Indeed attendance at meetings was small during the spring, sometimes so small that no meeting was held. Hooke was sometimes absent, and when present he was more interested in 'mechanical contrivances' than in experiments. Lower (once a possible Curator) seems to have lost interest in experiment after the publication of his *Tractatus de Corde*. In the summer Boyle was gravely ill, and, although he recovered to be active in the autumn, he does not seem to have attended meetings of the Society before the end of 1671, keeping in touch through Oldenburg and probably Hooke, presenting his works as they appeared and providing materials for several experiments made by Hooke before the Society in 1671. 1670 was, as a whole, a poor year for attendance, with several meetings being cancelled; the substance of those that were held was generally provided by Oldenburg's reading of the relevant parts of letters, many, like those of Edward Browne from Central Europe and Northeast Italy, being filled with technology and natural history.

Early in 1671 things seemed a little livelier: there were accounts of respiration experiments made with an 'air-vessel for a man to sit in' and projectile experiments; there was also eager discussion when Martin Lister's letters on plant physiology were read. (Lister, F.R.S. 1671, was a provincial physician with wide interests, including spiders and fossils.) But there was dissatisfaction also; on 9 February 1670/1

> It being observed, that very many things were begun at the Society, but very few of them prosecuted, Mr Oldenburg offered to bring in a list of such particulars, which he was desired to do with speed.

Such a list would have been long, and it would have been very troublesome to go through all the minutes of past meetings; equally, it would have been a great vexation to those many Fellows who had failed to bring in the experiments desired to be named as delinquent. It was industrious of Oldenburg to suggest the undertaking and hardly surprising that he, like so many others, failed to prosecute his task. On the whole, 1671 was a poor year for experiments, although somewhat less so than the previous year.

The opening of the year 1672 can have seemed little different, and indeed historians[28] have seen it as a year of still declining activity. The appointment of Nehemiah Grew as Curator for the anatomy of plants for the period of one year has been taken variously as a response to lack of activity and/or as an encouragement to Grew for his already well-advanced work on plant anatomy.[29] Whatever the motives, one at least must have been the recognition of Grew's excellence in experimental botany, coupled with the belief that both Grew and the Society would benefit from his presence in London and his more frequent attendance at Society meetings. There may even have been a touch of xenophobia (at worst) or merely of national pride involved, for at this time the Society, through Oldenburg, was encouraging Malpighi (F.R.S. 1668, three years before Grew) to pursue plant anatomy. (In the event, the Society sponsored books by both men and, unusually in this period, there was no rivalry shown by either.)

Between 1671 and 1675 the Royal Society's contributions to research on plant anatomy and to Malpighi's research on embryology were to be of great importance. It is thus of less long-range importance or significance that the Society did not, in the years 1672–4, witness the number of experiments which was its custom, far fewer than in the early years, for the meetings were well supplied with accounts of experiments performed privately by Fellows like Malpighi and Grew, accounts which were discussed at length and encouraged by correspondence. One may even hazard the guess that it was becoming realised that experiments on living matter were affairs of some delicacy: they were not easy to perform with a crowd of onlookers, who in any case could not all see what was happening, being difficult to perform successfully in any case, and often it required a trained eye to see what was to be seen. And, as noted above, when it came to physiological experiments on living animals there were some, like Hooke and, sometimes, Boyle, who developed a sensitivity to the moral question of whether it was desirable to repeat such experiments for the entertainment of the curious, desirable as it might be to perform them in the pursuit of knowledge.

Physical experiments were in every way more satisfactory, and generally exempt from the majority of the problems which arose in biological experimentation. And 1672 opened with two highly sig-nificant events related to the development of physical experiment. Isaac Newton, Lucasian Professor of Mathematics at Cambridge in succes-sion to Isaac Barrow, and known to a few of the Fellows as a promising mathematician, was elected a Fellow on 11 January 1671/2, having been formally proposed by Seth Ward, Bishop of Salisbury, on 21 December 1671, and his 'improvement of telescopes' was discussed at the same meeting. The Society had long been interested in practical optics: besides burning glasses, much discussed in 1665, there had been great interest in Francis Smethwick's (F. R. S. 1667) attempts to grind non-spherical lenses and in John Beale's account of optic tubes, both in 1666. Now the Society welcomed eagerly Newton's proposition and his demon-stration of the possibilities of reflecting telescopes. It is probable that Ward's proposal of Newton's election stemmed from the Society's receipt of one of his small reflecting telescopes which was seen by the King and then 'considered' by Brouncker, Moray, Neile, Wren and Hooke. These jointly decided that a description and drawing should be sent to Huygens at Paris, believing that publicity of this kind would be the

meanes to secure the Invention from ye Usurpation of forreiners [and]
prevent the arrogation of such strangers as may perhaps have seen it here,
or even with you at Cambridge,

as Oldenburg told Newton.[30] (Oldenburg mentioned to Huygens the
telescope's superior qualities in a letter of 1 January 1671/2 and sent the
detailed description with diagram a fortnight later.) The Society
examined two telescopes at successive meetings and discussed Newton's
advice about speculum metal;[31] when the larger telescope was discussed
for the second time (1 February), Hooke as commonly was asked to see
whether he could 'perfect it'. Meanwhile, since Newton had offered to
communicate an account of what he called the 'Philosophical discovery'
which had led him to attempt a reflecting telescope in the first place,
Oldenburg encouraged him to send to the Society what became his
letter of 6 February (read to the Society two days later) describing his
prismatic experiments for the first time, with his conclusion that 'Light
itself is a Heterogeneous mixture of differently refrangible rays'. As
Oldenburg told Newton in reply, its reading 'met both with a singular
attention and an uncommon applause'; thanks and a unanimous vote
that, with Newton's concurrence, to preserve his priority and to secure a
wider audience, it should 'without delay be printed', as it was in the
Philosophical Transactions for 19 January 1671/2.

It is not at all surprising that the Society as a whole was highly
enthusiastic, for Newton's paper displayed every attribute of natural
philosophy that the Society professed most to value: it was thoroughly
experimental, the experiments were ingeniously designed and appeared
to be clearly and carefully described, and the novel and remarkable
conclusion could clearly be drawn from these experiments. Or, as
Newton insisted, his conclusion was 'not an Hypothesis, but most rigid
consequence ... evinced by the mediation of experiments'. Further, the
paper offered a wealth of material to discuss and many experiments
suitable for repetition, and new experiments could be readily devised in
the light of Newton's 'doctrine'. Most Fellows who heard or read the
paper found it wholly admirable. Only Hooke demurred, because he
wholly rejected Newton's conclusions, having his own previously formulated hypothesis about light and colour which he maintained to be
consonant with Newton's experiments. In a sharply worded critique,
Hooke accepted Newton's experiments as valid but denied Newton's
conclusions, as others were to do. (Newton was in fact inclined to agree
that several *hypotheses* could be reconciled with his experiments, but he

argued that his conclusions or doctrine were not hypothetical, since the *doctrine* emerged directly from the experiments, which made it certainly true.) The Society as a whole was prepared to spend much time on Newton's interesting and even exciting experiments, requesting Hooke to repeat some of the more novel. This Hooke was reluctant to do. But some were soon repeated before the meetings and Hooke showed an experiment with a soap bubble[32] 'wherein there appeared something of water which had neither reflection nor refraction and yet was diaphanous'. Altogether the reading of Newton's papers, commentary and objections by Hooke, Huygens and Pardies in Paris in reaction to the publication of Newton's paper in the *Philosophical Transactions* (only Pardies was ultimately to be convinced of the correctness of Newton's conclusions) occupied a good part of some fifteen out of the nineteen meetings of the Society between early February and the July recess, while the Society would gladly have witnessed more repetition of Newton's experiments had Hooke been willing to provide them.

During the summer recess (the last meeting was on 10 July and the Society resumed its meetings at the end of October), Hooke had clearly been interested in other things, for when the Society reconvened he spoke of other experiments: that of the anomalous suspension of mercury in a Torricellian tube;[33] that of the attractive power of a sulphur ball when rubbed, as described by Von Guericke whose book he had been reading;[34] the effect of combustion on air; and the construction of his lens-grinding machine. Not for another month,[35] even with urging from the Society, did he read his own discourse 'containing divers optical trials', which, he believed, demonstrated 'some new properties of light'; this aroused nothing like the interest which Newton's papers had done, and to Hooke's undoubted disgust the Society valued Newton's contributions more highly.

Between Newton and Hooke, physical experiment had clearly dominated proceedings in 1672. Nevertheless, biological subjects had not been neglected. Lister continued to attract interest with his letters on plant anatomy, on which Grew commented when they were read to the Society on half a dozen occasions,[36] while Grew's activities as Curator were so well appreciated by the Society that (18 December) the Council approved the continuation of his appointment, providing that money to pay him could be found. (In fact, Grew soon returned to his practice as a physician in Coventry for some months only, before once again coming to London.)

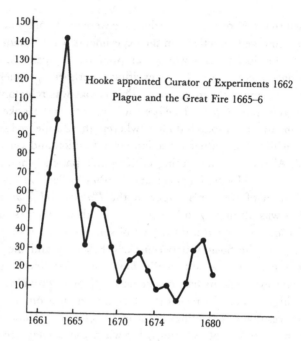

Figure 2.   Experiments performed annually at meetings, 1661–80.

In Grew's case, the post of Curator did not so much involve the performance of experiments before the Society, either because he did not wish to show them or because they were unsuitable for public demonstration, as it did the presentation of research papers and lively contributions to discussion of the papers of others. And this was to be increasingly the trend. In 1673 more letters and discourses were read than in earlier years, and many instruments and mechanical devices were produced and described to the detriment of experimental demonstration. Although Henry Hunt arrived to be Hooke's assistant,[37] Hooke's chief interest this year was in invention, not experiment. So when Leibniz's arithmetical machine was shown at a meeting, Hooke spoke of his own invention in that line, which he also showed; he produced a partially polished objective speculum of large curvature (36 feet), gave a discourse on a weathercock of his invention, showed a small fountain, a very small Leeuwenhoek-type microscope and a (not very satisfactory) apparatus for measuring 'the force of the loadstone's attraction at a certain distance'. His true experimental activity during this year was mainly quasi-chemical: on the production of 'air' in effervescence; on the loss of air in some cases of combustion; on two fluids which when mixed occupied less volume than separately. All these were reminiscent of past experiments published by Boyle, who himself provided several experiments during this year.[38]

After the anniversary meeting of 1673 with its election of several new Council members (including Boyle, Petty and Wren), there were new attempts to regenerate the Society, both by collecting arrears of subscriptions and by increasing the experimental content of the meetings. A committee was appointed (22 December) which was

> desired to draw up a list of considerable experiments to be tried before the Society, and to prepare any apparatus necessary for the exhibition of them upon all occasions.

Petty was 'desired to take a particular care of seeing the import of this order put into effect'.

As so often, it was easier to propose activity than to see that it was carried out, and during 1674 there were several weeks when the Society did not meet, and the meetings were adjourned early for the summer vacation. Hooke was particularly interested in astronomy during this year:[39] he showed an experiment to demonstrate the impossibility of making astronomical observations accurate to less than one minute of arc; he described a new sighting instrument which he had devised; he produced a Gregorian reflecting telescope of his own design, a small quadrant with telescopic sights and other instruments. He did perform a number of experiments, mostly on magnetism, with a view to substantiating a theory which, he believed, would permit the finding of magnetic variation anywhere on the globe. Grew showed microscopical observations regularly and, after Lister sent a specimen of a 'blood-staunching liquor', his styptic was tried on a dog with apparent success.[40]

But the meetings as a whole lacked vigour and variety as is evident from the minutes. As already described (Chapter 2), there was a strong sense of dissatisfaction among the Council and a determination to restore the activities of the meetings to something like the original vigour and liveliness. The opportunity to do so was made easier by the fact that in 1673 Gresham College had invited the Society to resume the holding of meetings on its premises, offering to provide a meeting room, a library and room for the repository. The move, after due consideration, was to be made in November 1674.

~ 4 ~

The communication of experiment
1660 ~ 1677

Besides the performance and the description of experiments at meetings, there were other ways, both formal and informal, by which the Royal Society and its Fellows sought to promote the Society's experimental aims and ideals. There was the writing and publication of works wherein the Society's aims might be praised, exemplified, interpreted and urged as models to be followed, and this on many levels. There was the communication to those, at home and abroad, not able to attend meetings, of what was being done at those meetings, as likewise the solicitation of exchanges of discoveries, experiments, inventions and views from such absentee members. And there was the encouragement of the fruits of experimental learning by the sponsoring of publication, in many and various forms.

That the Fellows constantly tried to exemplify the aims and ideals of the Society in their writings is very well known but worth stressing again. Of all the earliest Fellows, none was so zealous or so successful in this regard as Robert Boyle, who, as a virtually unknown natural philosopher, blazed into fame in 1660 as the author of a most remarkably experimental treatise, *New Experiments, Physico-Mechanical, Touching the Spring of the Air and its Effects*. It was topical, it was innovative and it was shot through and through with the author's profound belief in the value of experiment for discovery and, implicitly, for the testing of theory. This major work was to be followed in the next year by two more important books, *Certain Physiological Essays*, in which Boyle discussed the philosophy of experiment as well as giving original examples of its use, and *The Sceptical Chymist*, in which he continued to display the fruits of his chemical investigations, as he had begun to do in the previous work, in the study of the structure of matter and its properties. From this time onwards, there were few years in which Boyle did not produce either a treatise embodying his views on the importance of experiment (as *The Usefulnesse of Experimental Naturall Philosophy*, 1663 and 1671) or one embodying examples of experimental investigations, whether in the exploration of a particular area of the natural world (as *Experiments and Considerations Touching Colours*, 1664, or *The Aerial and Icy Noctiluca*, 1680–1) or embodying experimental arguments for his rejection of Aristotelian and Spagyrical elements and his promotion of his own 'corpuscular philosophy' (as in *The Origine of Formes and Qualities*, 1666) or a combination of all of these approaches. Boyle's writings were the most striking possible propaganda for the Royal Society's aims, for what Boyle in his first book called 'my grand design of promoting experimental and useful philosophy',[1] thereby exemplifying what he and others had been doing in Oxford before the Restoration and the purposes they had in mind in devising a formal London-based organisation. To foreigners particularly Boyle embodied the experimental aims of the Royal Society as they understood it, and until his death in 1691 he was almost universally held to be its greatest luminary and the best exemplar of these aims.

To the Society itself, Hooke also held an important place as a major practitioner of the experimental approach, partly on account of his office as Curator of Experiments, partly on account of his *Micrographia* of 1665. This was licensed by the Society's Council; the title page firmly announced Hooke to be 'Fellow of the Royal Society' (as Boyle's, for example, never did); and Hooke tactfully dedicated the work both to the King and to his Society. It was a striking and admirable piece of propaganda for the Society's aims, while Hooke in his Preface expressly claims to have 'undertaken [his work] in the prosecution of the Design which the Royal Society has propos'd to it self'. Hooke's work is of course far from being only a series of microscopical observations, excellent though these and the illustrations of them are. It contains as well a whole system (not it must be admitted very logically set out) of natural philosophy, most strikingly of optics, presumptively derived from his observations and experiments, and thus a good illustration of the Society's aims. Excellently conceived and contrived though these were, Hooke, devout Fellow of the Society, Boyle's former assistant and disciple, was never to show himself as thoroughly committed to the experimental way advocated by Boyle as was his master. Hooke was often attracted by an hypothesis once it fitted experiment, without considering whether it was the only tenable hypothesis and once he had formed such an hypothesis he was reluctant to abandon it, as his controversy with Newton in 1673 was to show. Clearly he never whole-heartedly accepted Boyle's view (which was that of Newton also) that experiment could and should test an hypothesis so thoroughly that the hypothesis must stand or fall by experiment. In fact few, other than Boyle and Newton, as yet did hold that any hypothesis could be proved by experiment. Hooke was by no means singular among the Fellows of the Society in this regard. But only those who attended meetings and knew of his private views would perceive this; to the reading public Hooke was a thoroughgoing experimentalist and certainly all his works were filled with excellent examples.

Sprat's *History of the Royal Society* of 1667 has already been considered above (Chapter 2). It was intended to be a propaganda piece, a rôle it amply fulfilled. A point to recall is that Sprat gave examples of some of the achievements of the early Fellows as he or more probably Wilkins saw them; many of these are unfortunately rather vaguely described or minor in importance, but all were experimental and therefore propaganda for the Society's views.[2] Sprat's *History* thus gave a not very accurate view of the Society, one which over-emphasised the importance placed upon experiment and empiricism, but all the more emphasising the method which Sprat intended to convey. As it had considerable influence at home and abroad – it was translated into French in 1669, while it went through three English editions in the eighteenth century (1702, 1722 and 1734) – it very much helped to maintain the view that the Royal Society was deeply imbued with Baconianism and, *a fortiori*, with experimental performance.

Another author who promulgated the Society's views, this time unofficially and as he saw them, was Joseph Glanvill. It is doubtful how much he knew of its aims and performance in 1661 when his earliest work, *The Vanity of Dogmatizing*, appeared, but in it Glanvill, like Sprat later, was concerned to defend the use of reason in religion, reason to him including the study of nature by natural philosophers such as Galileo, Descartes, Gassendi and Harvey. A revised version of this book published three years later, the *Scepsis Scientifica*, contained a preface addressed to the Royal Society, identifying the views contained in it with the Society's preference for studying nature without preconceived hypothesis; it was no doubt this work which led to his election as F.R.S. in the same year. In 1668 his *Plus Ultra* (with explicit encouragement from the Society) argued that the Society's experimental philosophy was doing more to advance learning in a few years than had been accomplished in as many centuries by dogmatic philosophers, Boyle receiving especial praise in this regard. Later still, Glanvill was to defend the Society against the overt attacks of Henry Stubbe in a series of pamphlets, but this had little to do with the question of the value of experiment as such (see below, Chapter 9). Glanvill's views were not identical with those of other Fellows, and in particular he was more inclined towards a whole-hearted acceptance of the reality of witchcraft than were many others, but at least in so doing he endeavoured to subject all accounts to empirical tests as well as he could. For he truly believed that the investigation of witchcraft was comparable to the study of other natural phenomena as well as being a useful aid to religion. In sum, he was a minor but at the same time a useful propagandist for the experimental way.

These writers all believed in and tried to exemplify the Royal Society's point of view and by their works advanced the Society's experimental aims, directly or indirectly, by precept and example. More closely connected with the Society itself was the work of successive Secretaries, for whom explicit provision had been made under the Charters.[3] In these, two men were named, John Wilkins and Henry Oldenburg. The first served until 1668, the second until his death in 1677. It is clear that Wilkins and his successors left the main burden of the office to Oldenburg, neither attempting to be present at all meetings to take the minutes, nor overseeing the amanuensis, nor maintaining official papers – Journal Books, Council Books and the Register – nor drawing up the letters to be written in the name of the Society and Council, all duties specified under that Statures of 1663 and often referred to in the minutes. But they did make other contributions. All served on the Council during their terms of office. The contribution made by Wilkins in overseeing Sprat's *History* has already been mentioned. He was succeeded as Secretary by Thomas Henshaw, who served from 1668 to 1672 and from 1675 to 1677; his most useful work was probably that done not as Secretary, but (1672–5) while serving as envoy extraordinary to Denmark. From Copenhagen he sent back many letters on the climate and natural history of Denmark and, more importantly, brought within the Society's orbit Thomas and Erasmus Bartholin, the latter of whom particularly contributed to the Society's interests. During part of the time that Henshaw was abroad (the first year) John Evelyn served as Secretary. Like Henshaw an Original Fellow, he had been active since the beginning of the Society, mainly in publishing numerous books relating to the Society's early interest in the history of trades: after *Fumifugium* in 1661, an early tract on air pollution, came *Sculptura* in 1662, on copper engraving, *Sylva*, on forestry, with *Pomoma* on cider both in 1664, the first works to be licensed by the Society and in fact compiled (for Evelyn was principally editor) at the Society's express direction,[4] and numerous works, both original and translated from the French, on gardening. Evelyn was succeeded for the remainder of Henshaw's time abroad by Abraham Hill, one of the founder-organisers of the Society, chiefly an administrator, Treasurer 1663–75, 1677–99 and serving on many committees, especially those concerned with natural history. His main contributions to the Society's experimental programme had been made earlier through correspondence: as his *Familiar Letters* (published in the eighteenth century) show, he wrote frequently in 1663 to a country member, John Brooke of York, F.R.S. 1662, always about experiment and invention, sending him copies of many of the 'discourses' read to the Society during this year, and between 1663 and 1669 he engaged in correspondence with Nicholas Witte, physician of Riga, to whom he sent English books on natural philosophy.[5]

None of these Secretaries contributed more than minimally during their terms of office to another, very positive, method of promoting the Society's aims: this was the improving of experimental philosophy by encouraging others and by learning what was done elsewhere through the medium of correspondence. Yet the need for this had always been envisaged. As early as the first organisational meeting on 28 November 1660 it was decided to appoint William Croone as 'register'. Presumably the original intention was simply that he should keep the minutes and record any experiments performed and so on, for at the next meeting he was instructed to find an amanuensis who knew shorthand to assist him in his duties. Quite soon, however, he began to correspond with provincial natural philosophers, thereby to bring them into the orbit of the London group, either by informing them of their election and the duties required of members in promoting experimental philosophy, or by urging them to send information about their own experimental activity. An excellent example of how correspondence might bring this about and incidentally stimulate a man to greater activity is the case of Henry Power, physician of Halifax. Croone wrote to him on 20 July 1661 asking for information about the magnetic and pneumatic experiments at which he was said to be at work.[6] Encouraged by Croone acting on behalf of the Society, Power over the next two years, as noted above

(Chapter 3), sent detailed accounts of these and many other experiments, some undertaken independently, some performed at the Society's suggestion. And these culminated in the account of his microscopical investigations, which must have stimulated Hooke to continue his own. The nature of Power's experiments is to be seen in his *Experimental Philosophy* (London, 1664, but printed off in 1663). Rather curiously, Power gave no indication whatsoever in the printed work of the encouragement he had received from the Royal Society, even when describing experiments suggested to him in correspondence, nor is there any specific mention of the Society in his concluding dithyramb in praise of the rise of experimental philosophy in the mid-seventeenth century:

> This is the Age wherein (me-thinks) Philosophy comes in with a Spring-tide ... These are the days that must lay a new Foundation of a more magnificent Philosophy, never to be overthrown: that will Empirically and Sensibly canvass the *Phaenomena* of Nature ...

By the summer of 1661, when Croone first wrote to Power, plans were already underway to obtain a formal Royal Charter of Incorporation. Two months later the text was agreed upon, to be presented to the King in October, and granted on 15 July 1662. This First Charter had, as an important feature, not only the right to appoint two Secretaries, but even more importantly

> full power and authority, by letters or epistles ... to enjoy mutual intelligence and knowledge with all and all manner of strangers and foreigners, whether private or collegiate, corporate or politic, without any molestation, interruption, or disturbance whatsoever,

the only limitation being that the correspondence must be for 'the particular benefit and interest ... of the Royal Society in matters or things philosophical, mathematical, or mechanical'. This generous provision meant much in time of war (in fact, during most of the 1660s and 1670s), and made it clear that the Society was able to communicate with foreigners as freely as with Englishmen. Freedom of communication meant openness, absence of secrecy in a way uncommon before this time, and not by any means altogether normal even in the 1660s. The Accademia del Cimento, for example, was a small closed group, while the Académie Royale des Sciences neither admitted outsiders to its meetings nor had any regular means of communication or publication; its members either corresponded and published as private individuals, or they published cooperatively, or anonymously and in the name of the Académie. This privilege also meant that the Royal Society's official aims and its dedication to experiment were to be widely disseminated throughout the learned world, without, of course, demanding internationalism in the broader sense – for the Fellows of the Society were often moved to defense of English priority at best and to positive xenophobia at worst.

It was Oldenburg who was to take on almost single-handedly this extremely important aspect of the Society's affairs.[7] At first he had assistance, for although Croone ceased to be active in this way after the creation of the post of Secretary under the Charters, others continued regular unofficial correspondence, especially useful when with foreigners. Thus Sir Robert Moray, who had long corresponded with Christiaan Huygens, continued to do so until 1669; Oldenburg, who had met Huygens in 1661 and then corresponded with him privately, only took up regular official correspondence with him, at Moray's instigation, in 1665. Wallis had corresponded for many years with a number of foreign mathematicians and astronomers, notably in the 1650s, but after 1663 such letters mainly went through Oldenburg, who for example took up a correspondence with Johann Hevelius (formerly a correspondent of Wallis's) in 1663. But Wallis and Moray and others had exchanged news and information with several different correspondents only as private individuals and without any attempt, as Oldenburg was to do, to establish a genuine network of correspondence. Oldenburg both served as a distribution centre for news and acted to stimulate individuals to discuss, argue and reveal their own work and points of view. And this was so much the more important because Oldenburg, though very often, in fact, acting on his own initiative, could claim to be acting as the representative of the Society, and was so viewed by his correspondents, so much so that in years to come foreigners very often saw him as embodying everything that the Society stood for *in propria persona*. In 1668, hoping to persuade the Council to give him some assistance, he described his secretarial work in detail; relevant here is the latter part of his plea:[8]

> He [the Secretary] ... writes all Letters abroad and answers the returns made to them, entertaining a correspondence with at least 30. persons; employs a great deal of time, and takes much pains in inquiring after and satisfying forrain demands about philosophical matters, dispenseth farr and near store of directions and inquiries for the society's purpose, and sees them well recommended etc.

(Later the number of his regular correspondents was to increase.) That his efforts were understood and appreciated by others is apparent from the unsolicited tributes of very many of his correspondents. One of these, Joseph Glanvill, went further and praised him in print, declaring[9]

> Mr Oldenburgh, Secretary of the Royal Society ... also renders himself a great Benefactor to Mankind, by his affectionate care, and indefatigable diligence and endeavours, in the maintaining Philosophical Intelligence, and promoting Philosophy.

It was Oldenburg's practice both to instigate correspondence – as when first writing to Hevelius or later to Marcello Malpighi of Bologna inviting a return – and to respond favourably to overtures from foreigners who aspired to be recognised by the Royal Society, provided they seemed to show the correct spirit and interests. In both cases he did his best to expound that philosophy of the Society, emphasising above all its reliance of experiment as against theory, its encouragement of communication of discoveries to the world at large, and its desire for the accumulation of knowledge about the natural world. An excellent example is contained in letters written in 1670 to the young, unknown Leibniz and to his patron, Baron von Boineburg.[10] In that to Leibniz he spoke of 'those who have it much to heart that the wise and industrious of all nations should combine their studies and efforts towards the increase and perfection of a solid and fertile philosophy', while to Boineburg, a distinguished diplomatist, he exclaimed

> Would that those who excel in litigation and in the sciences in our Germany would make their contributions towards the restoration and perfection of philosophy with a better will than they have shown hitherto, and would eagerly imitate in this the example of England, France and Italy itself in turning to experiments.

Oldenburg, as here and elsewhere, always endeavoured to cast the net of the Society's ideals and influence over all European nations. Usually, however, he did so with distinctly English patriotism, not to say chauvinism, although he always retained a remembrance of his German origins.

But more than all else, Oldenburg strove to stimulate selected natural philosophers to complete and communicate their experimental result, whatever their nationality, whatever their standing, from the great Huygens with his international reputation to the young Newton known only obscurely as a Cambridge mathematician. Moreover, he often elicited information from one man by telling him what another was doing and even encouraging controversy. For like Mersenne before him (but Oldenburg could not have known directly of his network of correspondence), he believed that a certain amount of acerbity was the most likely way to stimulate those reluctant to communicate their results. And, more often than not, the method worked, and usually, because of Oldenburg's epistolary tact, without real acrimony. Thus between 1665 and 1667 Oldenburg mediated between and provoked the resultant participants to communication in a prolonged controversy over cometary observation and theory involving exchanges between both Auzout and Hooke and Auzout and Hevelius. In the first exchange he was decidedly on the side of Hooke.[11] Later, in 1673–4, when Hooke vehemently attacked Hevelius for his neglect of telescopic sights, he was able to steer a middle course, guided by the advice of Wallis, who valued Hevelius's experience in observation over Hooke's necessarily better instruments.[12] In 1673 also, Oldenburg was forced to mediate between Hooke and Newton after Hooke's ill-mannered attack on the younger man's first optical paper, and was soon to persuade Newton to reply at length to a stream of criticism from abroad, criticism which renewed itself in 1675 after Newton's second paper; his critics included Huygens and Pardies in Paris and Line and his pupils in the Low Countries, and the whole affair was managed without acrimony on the part of any of the correspondents, including Newton, who then hardly complained.[13] It should be noted that in this case these were not priority disputes, which often did engender much heat (although Newton was at this time quite happy to concede Sluse's independence in developing 'the method of tangents'),[14] but controversy over experimental methods, results and interpretation. And although, obviously, these could have been as acrimonious as priority disputes, the fact remains that they were not, and this is thanks in good part to Oldenburg's skill. After Oldenburg's death, Newton quickly withdrew from his optical correspondence, refusing to carry it on now that he would not disoblige Oldenburg by ceasing.[15]

Besides the implementation of the Royal Society's experimental ideals through correspondence, other modes were open to it and its Secretaries. Just as Wilkins had guided Sprat in presenting the Society's ideals in print, publication of other men's work was used by Oldenburg to the same end. Fostering of publication demonstrated clearly the Society's ideals, and in this Oldenburg, mostly under the Society's aegis, was very active. The most important example is that of the works of Malpighi, virtually all after 1669 published in London by order of the Society, to whom they were dedicated, with those printed before 1677 all edited by Oldenburg, that is assembled and seen through the press by him; they were partly in the form of letters sent to Oldenburg.[16] Less officially, Oldenburg used his linguistic skill to translate and have published English versions of a number of foreign works, both Latin and French: examples of those relevant to the spread of experiment were Steno's *Prodromus* (1669; English translation, London, 1671) and Moise Charas's *Nouvelle Experiences sur la Vipère* (Paris, 1669; englished as *New Experiments upon Vipers*, London, 1670).

But Oldenburg's most important venture into publication was, without any doubt, his foundation in 1665 of the *Philosophical Transactions*. This private journal (as it long remained) he carefully entitled *Philosophical Transactions: Giving some Accompt [sic] of the Present Undertakings, Studies, and Labours of the Ingenious in many Considerable Parts of the World*, as the title pages of all volumes read until that of volume 66 (1766).[17] For this venture Oldenburg naturally enough sought (and secured) the Society's approval and its imprimatur, in return for which the Council demanded the right of scrutiny; as the Council Minute of 1 March 1664/5 runs

> That the Philosophical Transactions, to be composed by Mr. Oldenburg, be printed the first Munday of every moneth, if he have sufficient matter for it; and that the Tract be licensed by the Council of the Society, being first reviewed by some of the Members of the same; and that the President be desired now to license the first papers thereof.

For the next couple of years a certain amount of supervision was maintained. Thus in 1665, when plague made London unsafe, numbers 6, 7 and 8 (for November and December 1665 and January 1665/6) were printed at Oxford, where Wallis, Moray and Boyle all took some editorial part.[18] John Beale from Somerset (but he was never on the Council) kept a close eye upon much that Oldenburg wrote, especially in regard to agricultural and horticultural matters, offering many long and detailed comments upon the contents, comments which Oldenburg often printed, while he sometimes submitted his prefaces to each volume to Beale for approval. But generally, as the years went by, Oldenburg was given a free hand by the Council, and it was always made clear, as far as possible, that the work was Oldenburg's own. Yet many readers would persist in believing otherwise, so much so that in May 1666, at the

end of issue number 12, Oldenburg felt compelled to insert an advertisement:

> Whereas 'tis taken notice of, that several persons perswade themselves,
> that these *Philosophical Transactions* are publish't by the Royal Society ...
> The Writer ... hath thought fit, expresly here to declare, that that
> persuasion, if there be any such, indeed, is a meer mistake ...

Foreigners were particularly prone to make this mistake. Oldenburg
was seriously annoyed to find the 1672 Latin version of the volumes for
1666–9 published with the title *Acta Philosophica Societatis Regiae in Anglia
... auctore Henrico Oldenburgio,* and complained bitterly to the translator.[19] Even the *Journal des Sçavans,* reviewing the first issue, had been
rather vague, not, it is true, specifically attributing it to the Royal
Society, but not to Oldenburg either.[20] It was the *Journal* itself that
began the myth that it truly antedates the *Philosophical Transactions* as a
scientific journal. To be fair, the phrase there used is 'un Journal, pour
apprendre aux Sçavans ce qui se passe de nouveau dans la République
des Lettres', which it did by reviewing all new learned books, from
theology and law to literature and natural philosophy and later by
occasionally publishing short accounts of scientific discoveries, particularly those by Huygens (sometimes when Oldenburg suggested it!).
But Oldenburg's plan was different. He did print reviews of books, but
normally only of books related to natural and usually experimental or
empirical philosophy. But his specific aim, as he said in the introduction
to the first issue, was 'promoting the improvement of Philosophical
Matters', for which he thought 'nothing more necessary ... than the
communication to such, as apply their Studies and Endeavors that
Way, such things as are discovered or put in practice by others'. It was
thus not at all on the same plan as the *Journal des Sçavans,* and it seems
only right to give Oldenburg credit for establishing the first journal
primarily devoted to what was new in the world of natural philosophy,
usually before its authors could write an extended account.

The first number contained accounts of inventions and discoveries
derived partly from his own knowledge, partly from accounts read to the
Royal Society (by no means all from English sources), partly from
letters, and partly from printed sources. By the second number,
Oldenburg was settling into what became his pattern: extracts of letters,
English and foreign, the latter translated into English unless in Latin
(especially astronomical and mathematical letters were so left) and, at
the end, one or more book reviews (e.g. Hooke's *Micrographia* was

reviewed in issue number 2); after issue number 10 (March 1665/6) this section was listed in the table of contents as 'An Account of some Books, lately published'. The majority of these reviews were by Oldenburg, although Wallis helped with mathematical books and so probably did some other Fellows. Soon the more active Fellows, like Wallis and Boyle, many country virtuosi, like Joshua Childrey and Nathaniel Fairfax, and many provincial Fellows, like Martin Lister, became accustomed to writing letters which were really essays intended both for reading to the Society and for publication – such as Newton's first 'letter' on light and colours on 1672. But never during Oldenburg's editorship was the material for the *Philosophical Transactions* drawn exclusively from the Royal Society's proceedings, nor from purely English sources. The journal remained, as he intended, an account of 'the Labours of the Ingenious in many Considerable Parts of the World' and contained always a large amount of experimental material, together with natural history, while mathematics, and especially astronomical observations, were also included. Long before their experimental results were available in book form, Wallis, Boyle, Lister, Newton, Leeuwenhoek, and even sometimes Huygens, were able to make them known to the learned world through the letters they wrote to Oldenburg, which might be published whether or not they were read at meetings (although they usually were). It was of an inestimable advantage for the communication of the Royal Society's experimental ideals and practices, and to a considerable extent brought such work by European savants and natural philosophers within the aegis and sometimes patronage of the Royal Society.

The *Philosophical Transactions* were immensely popular with all, whether English or foreign, who could not regularly attend meetings, and with most who could. In spite of the fact that most accounts were in English, foreigners everywhere welcomed them and wrote to beg for copies to be sent by any means available (they were too bulky for the ordinary letter post). All could look at illustrations and diagrams, some knew enough English to read at least some of the letterpress, even if with difficulty, and some fell back on the kindness of English-reading friends. The existence of more than one partial translation into French of some issues shows how much they were valued in France. Both the *Journal des Sçavans* and Jean Denis in his *Mémoires concernant les Arts & les Sciences* published occasional extracts, almost always of dramatic or startling discoveries (like Newton's reflecting telescope) or experimental advances, while Oldenburg of course drew on other journals for some foreign news. Most striking is the tribute of Francesco Nazari, founder of the Italian *Giornale de'Letterati* in 1668; this, like the *Journal des Sçavans*, aimed to cover the whole world of learning. He wrote in flowery compliment,

> When I undertook to put together a learned journal, straightway my thoughts took wing for London, as to a glorious market whence a very rich supply of all philosophical commodities is to be obtained,

describing the *Philosophical Transactions* as 'filled with so many and such important observations and experiments, that they may at length suffice for the establishment of a true philosophy'.[21] It is notable how often unknown foreigners writing to Oldenburg used their admiration of what they found in the *Philosophical Transactions* (which they might only know in Latin or French versions) as an introduction. The *Philosophical Transactions*, like his correspondence, were influential contributions by Oldenburg to the spread of the Royal Society's ideals in the 1660s and 1670s and were a great encouragement to experimental philosophy everywhere. After Oldenburg's death it was to take several decades to re-establish the journal properly; that natural philosophers everywhere felt that it was an essential part of intellectual life is one of Oldenburg's greatest tributes.

Oldenburg's death in 1677 marks the end of a stage in the Society's development for two reasons. First, he had been a remarkably able and conscientious secretary; others, from Wilkins through Hill, did little as joint Secretaries, although much as Fellows. It was Oldenburg who, single-handedly, managed the Society's correspondence, both domestic and foreign, took and kept the minutes of both Council and Ordinary Meetings, and in 1665 had conceived and begun publication of the *Philosophical Transactions* which, it must be remembered, was his private venture, however much it might be associated in the minds of its readers with the Society itself. (In the same way, Oldenburg was, to the public at large, so closely associated with the Society that all he said and did was taken to represent the will and views of the Society.) Oldenburg's virtues and defects both had a profound influence upon the conduct of the meetings. He was not in any sense a practising scientist and so could not add directly to the Society's experimental activities, the nearest he came being in gathering material for the universal natural history which he often spoke of as the Society's aim. His contributions to meetings lay in the collecting of information from others and the choosing of excerpts from letters to read. Further, such letters, with others which he did not read publicly, he usually sent to Fellows not resident in London: to Wallis in Oxford, to Boyle when he was absent from London or not able to attend meetings, to Newton in Cambridge, to Flamsteed in Derby and later Greenwich, to Lister in York, to Huygens and others in Paris, and to Hevelius in Danzig. And from these he expected and usually received comment, reply and extension of the subjects discussed. He conducted mathematical (with help usually from John Collins), physical, biological and experimental correspondence with skill and, on the whole, tact, so that complaint was unusual. The case of Hooke, who did complain, *was* unusual; whatever Newton may have thought and said later about Leibniz (and there Collins, as Newton must have known, supplied the information that Oldenburg transmitted), he never blamed Oldenburg for supplying mathematical information to Sluse or optical information to a host of critics, including Huygens, Pardies and Line. Oldenburg succeeded admirably not only in keeping Fellows informed but in persuading them to disclose scientific ideas and discoveries they possessed but had not made public and in turning their minds to problems they had not previously pursued deeply. When even Hooke could not always repeat Newton's experiments at meetings successfully, it is no wonder that foreigners, not so predisposed in his favour as fellow-members of the Society might be expected to be, doubted his conclusions because they could not reproduce his experiments. Here Oldenburg's careful transmission of such difficulties often stimulated Newton to more complete and more thoroughly worked out directions that made it possible to reproduce easily his experiments, no slight contribution to experimental science.

Secondly, Oldenburg was, also, on the whole, a faithful preserver of the Society's records. He took rough notes of all meetings, later writing them up carefully for copying into the Journal Book. There were very few meetings at which he was not present (except during his imprisonment in the Tower in the early summer of 1667)[22] until the last year of his life, when he was much distracted from his duties, apparently not entirely by his quarrel with Hooke. The record shows that he then sometimes took no minutes or failed to write them up. Hooke was to complain that Oldenburg had failed to give him credit for all his activities at meetings: the record shows that Oldenburg did not intend to slight him but left room for Hooke to supply details to his own satisfaction of the communications which he made, room which Hooke often failed to fill up. Of the foreigners towards whom, in Hooke's view, Oldenburg showed excessive partiality, both Huygens and Hevelius were as much Fellows as Hooke, and as there was then no distinction between domestic and foreign Fellows; the latter had every right to know what passed at meetings. It is true that in the case of Hooke's dispute with Hevelius the Society on the whole did not take Hooke's side, while in the case of his dispute with Huygens both Oldenburg and Brouncker were overtly partial. In the first case, Oldenburg could exercise no influence, although in the second he might, and certainly the Council, under Brouncker's leadership, defended Oldenburg publicly. But though partial, Oldenburg had acted properly throughout, betrayed no secrets and told Huygens nothing that was not public knowledge, while it was Hooke, not Oldenburg, who had influence at Court, Oldenburg's only Royal patron being Prince Rupert, not the King. When Hooke resented and disapproved of Oldenburg's activities, he was disapproving of the Society's right, specified in the Charters

> to enjoy mutual intelligence and affairs with all and all manner of strangers and foreigners ... [for] the particular benefit and interest of the ... Royal Society in matters or things philosophical, mathematical, or mechanical.

But in fact Hooke and presumably others were inclined to distrust all communication and openness in matters philosophical or mechanical, and in the autumn of 1674, when re-organisation of the conduct of the meetings was under discussion with a view to increased experimental content, the Council also discussed the whole question of whether it was better to have closed or open meetings. This subject was probably raised by Petty who, as noted above (Chapter 2), was very active in trying to obtain reform generally, for he was in the chair as Vice-President at the Council Meeting of 15 October 1674 when only Petty himself, Goddard, Colwall, Oldenburg and John Creed (a faithful attender but not a contributer to ordinary meetings) were present. Then, as the minutes report,

> It being represented, that the permitting of such, as are not of the Society, to be present at the meetings thereof, is both troublesome and prejudicial to the same, it was ordered, that the repeal of the statute, which allows such an admission ... shall be propounded at the next meeting of the council.
>
> It being likewise represented, that the liberty of divulging what is

brought in to the meetings of the Society is also prejudicial to the same, and renders divers of the members thereof very shy of presenting to them what they have discovered, invented, or contrived; it was moved, that a form of a statute might be prepared, injoining secresy to the members of the Society in such matters, as shall be brought in, and by the President or Vice-President declared to be kept secret, as the communicators desire.

At the next meeting, four days later, the statute permitting non-Fellows ('strangers') to attend meetings was noted upon and repealed. It is not at all apparent that their presence had affected the conduct of the meetings in any way, why it had been felt necessary to repeal the statute, nor, in the end, how well the prohibition was maintained, for from time to time in the future non-Fellows were to be invited to appear to make observations, show experiments or present information.

Clearly things had got progressively worse in the working of the Society in later years, especially during 1676 and 1677. In the latter year especially, Hooke was disgruntled, Oldenburg distracted, and attendance very poor. It was a sad ending to Oldenburg's fifteen years as Secretary. But more, it seemed to show that the Society, having lost its first vigour, needed great efforts to regenerate itself, to find the will to continue and younger men to suggest new paths to follow.

~ 5 ~

The record of the minutes
1674 ~ 1703

The problems faced by the Council when it met to determine future strategy in the autumn of 1674 have already been discussed (Chapter 2). To summarise, it was finally determined (7 October)

> That such of the Fellows, as regard the welfare of the Society, should be desired to oblige themselves to entertain the Society, either *per se* or *per alios*, once a year at least, with a philosophical discourse grounded upon experiments made or to be made

with a forfeit of five pounds in case of failure, a heavy fine when a year's subscription amounted to only a shilling a week. In spite of the Council's firm resolution, nothing was placed upon the statutes, and the new proposals were minatory only, never binding. But, as will be seen, they worked fairly well for the next year or two and set the pattern for what became the most usual mode of 'entertainment' at meetings, that is, the reading of papers, while experiments performed at meetings became fewer.

The meetings in the autumn of 1674 began as intended. Wallis read 'a discourse on gravity and gravitation grounded on experimental observations'; Boyle presented the Society with 'Experimental Notes of the mechanical production of Fixedness ... ', which Oldenburg read for him (it was to be published the next year); Petty read his 'Discourse concerning the Importance and Usefulness to Human Life of the Consideration of Duplicate and Subduplicate Proportion' (a somewhat curious work, not really an *experimental* discourse); while, after the anniversary meeting, Hooke read 'his discourse, concerning the construction and uses of his new quadrant';[1] Grew his 'Discourse concerning the nature, causes and power of mixture' on 10 December, when 'several of the experiments mentioned in this discourse were exhibited after it was read'; and, a week later, Oldenburg read two 'treatises' by Ray on botany. Most of these papers or discourses were long enough to

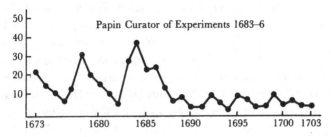

Figure 3. Experiments performed annually at meetings, 1673–1703.

occupy much of a meeting, while some of the longer ones were read over two or more.

Thus the new form of meeting seemed a viable one and set to continue, with (7 January 1674/5) a long experimental letter from Lister, effectively a discourse, being read to the meeting. A week later, those Council members present were asked to name the days when they would provide the entertainment, and the Earl of Ailesbury, Henshaw, Sir James Shaen, Brouncker, Pepys and (later) Southwell and Evelyn all volunteered, to provide for the next couple of months. (In fact Ailesbury, unexpectedly called out of town, Sir John Lowther (whose volunteering is not recorded) and Brouncker all compounded, that is paid the forfeit, and there is no record of Pepys and Shaen.) But other material was found: on 14 January 1674/5 Hooke 'read his observations of the late lunar eclipse', which he was asked to 'perfect' for publication, and Boyle's 'discourse of freezing' was read from the Register Book, and this provoked discussion with suggestions for further experiments. On 21 January Daniel Cox lectured on 'the analysis of vegetables', a chemical paper which aroused interest, and Hooke raised his views 'on the nature of trees, viz that it consists in the dissolutions of bodies by

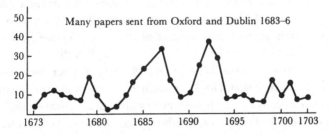

Figure 4. Experimental papers read annually at meetings, 1673–1703.

air', a view which was vigorously debated. For the rest of the year discourses continued: Hooke lectured on helioscopes and on lamps (these were Cutlerian Lectures and paid for not by the Society but by Sir John Cutler); Dr Edmond King read papers on physiology, especially muscular motion; Croone read a paper on flying (taking Ailesbury's place); Henshaw lectured on the natural history of Denmark (where he had resided for three years); Grew spoke on the anatomy of trees, on tastes, on the 'trunks' of plants, on the 'nervous liquor'; Dr Vossius spoke on the appearance of the moon, on Archimedes' burning glass and later on the motion of the seas and winds; Cox lectured again, this time on agriculture; Hooke again on light; Southwell on water; Needham on blood serum; while Boyle's *Discourse concerning the mechanical Production of Tastes* and Evelyn's *Philosophical Discourse of Earth* (usually known as *Terra*) were read probably by either Oldenburg or the amanuensis. On those occasions when no discourse was forthcoming, either one was drawn from the Register Book or letters were read, most notably, at the end of 1675 (9 December) Newton's discourse, his 'hypothesis of light and colours'. Also, especially in the latter half of the year, experiments were described or performed. Discussion seems to have been lively and the new procedure a success: although it produced few experiments actually performed at meetings, it did produce long and detailed accounts of experiments performed elsewhere. One which aroused special interest was Newton's experiment 'of glass rubbed to cause various motions in bits of paper underneath'; when first tried it failed – as the minutes charitably and cautiously record 'in the circumstances, with which it was tried' – but when tried subsequently it did succeed.[2] Here it seems to have been the case that the desire for repetition arose rather from curiosity, a wish to see the show, perhaps, rather than disbelief in the experiment as reported, and this was true of several of Newton's experiments.

Indeed, in this area at least Newton was an abler experimenter than Hooke, for several of Newton's experiments which both Linus and Hooke found difficult to repeat succeeded ultimately for the Society, although only after Newton had provided more explicit directions.

In 1676, Newton's hypothesis occupied three further meetings, provoking much discussion. Hooke performed a number of experiments on magnetism, always a suitable subject for public demonstration and of perennial interest. In February Oldenburg read an account of pneumatic experiments made in Paris by Huygens and Papin; some of them were repeated at the Society a week later (24 February). Grew furnished a total of five discourses in the course of the year, for which he was paid as promised, with expenses for the experiments he showed; presumably this was the result of the discovery in the spring that no one else was willing to volunteer the necessary experimental discourses.

Indeed on 6 March 1675/6 the Council became worried again and Brouncker as President

> moved, that it might be considered how to provide for the weekly meetings of the Society a sufficient number of experiments to be made from time to time, and to pitch upon such persons, as might be depended upon for the exhibiting of them.

Clearly Hooke's efforts were no longer enough. It was agreed that any Fellow willing to undertake making experimental discourses would be offered a maximum of four pounds plus 'the charges requisite to make the respective experiments', as was done with Grew. But most Fellows apparently did not find the offer of payment a sufficient stimulus, whatever those who drew up schemes to improve the Society's meetings as regards experiment might think. True, Newton's experiments were repeated and Croone described an experiment 'of exhausting the air out of spirit of wine' in order to estimate its effectiveness when used in a thermometer (for which he was presumably paid), yet meetings in the spring and early summer of 1676 were mainly occupied with letters. The Society adjourned in mid-June; even before this date the meetings were thinly attended (Hooke noted, 'Not 8 of the Society' in his *Diary* of 18 May) and Hooke himself was sometimes absent, and even when present was clearly not keen to devise and show experiments. He was still sulking over the support given by Brouncker and Oldenburg to Huygens's claim for priority in the application of spring balances to watches, a dispute which reached its climax with the printed postscript, highly defamatory, to *Lampas*, published in the autumn of 1676, directed against Oldenburg, whom the Council then defended. Hooke contented himself with sometimes reading a Cutlerian and/or a Gresham Lecture before the meeting, but this was not strictly an integral part of the Society's business, nor what he was paid by the Society to do.

At the beginning of 1677, as throughout the latter part of 1676, Hooke was still concerned with experiments on magnetism, correctly anxious to stress the dubiety of magnetic inclination as a means of determining longitude; these experiments supplemented those of others and were not particularly novel. It was always easy to devise and show such experiments, always difficult to give them any real theoretical significance. The fact that magnetism was a recurring subject for demonstrations performed at meetings throughout the years does not so much indicate continuity as the fact that no firm conclusions could be reached as to their meaning. Hence magnetism, like meteorology, remained an entirely empirical science. Meanwhile, Grew continued to read lectures regularly: several on salts found in various plants and their behaviour in solution, accompanied by chemical experiments, then two discourses on the comparative anatomy of animals, with particular reference to the stomach and guts, then one on the colours of plants. Many letters were read from foreign correspondents; in particular, six meetings were chiefly taken up with the reading of letters from Leeuwenhoek, whose microscopical work was to continue to interest the Society for many years and to provide sometimes much needed experimental content to the meetings. At this time there were also accounts of the recent discovery of various kinds of 'phosphorus', including true phosphorus (the element), which naturally aroused very considerable interest, and accounts of such astronomical phenomena as the comet of 1677.

It was in 1676 that Hooke began to plan his various clubs and societies, as discussed above (Chapter 2); that others at least temporarily fell in with his schemes suggests that they too were, as he was, dissatisfied with the workings of the Society.[3] Hooke laid all the blame upon Oldenburg, who was indeed much distracted by this quarrel. And while the fewness of the experimental presentations may be laid at Hooke's door, the occasional lacunae in the minutes between 1675 and 1677 and the lack of a Letter Book for the first half of 1677 must result from Oldenburg's laxness, although the failure to publish any number of the *Philosophical Transactions* in July and August arose because the printers were on holiday. Oldenburg's death in September 1677 marks an administrative hiatus in the Society's development but not any very sudden change in the content of meetings such as had occurred in 1674. The greatest loss was the decline of foreign contributions; Oldenburg, as noted in Chapter 4, had been an indefatigable foreign secretary, eliciting much information, much of it experimental, and managing complex interchanges between foreigners and Fellows with tact and skill. No one ever replaced him wholly, although things improved when Sloane was Secretary. Now some foreign correspondence was to continue – notably in letters from Leeuwenhoek whose only outlet was the Royal Society – but otherwise the flow of letters from the Continent thinned for some time.

Hooke clearly intended to use his opportunity to introduce a revolution in the Society's affairs. As his *Diary* shows, his first step was to search Oldenburg's papers for proof of his 'treachery' (that is, his spreading news of what occurred at Society meetings, especially in relation to Hooke's activities), his next to try to stop any such thing for the future. As already noted, steps had been taken three years earlier to try to prevent the presence of 'strangers' at meetings (which the statutes had allowed subject to the President's permission) and the relevant statute had been repealed (9 October 1674). The repeal had never been strictly applied, but now and for many years to come there is no record of 'strangers' being present.

The first autumn meeting (18 October 1677) saw something of the administrative shape of things to come: Brouncker did not attend and Hooke was invited by those present to take Oldenburg's place. At the Anniversary Meeting, Hooke was confirmed as Secretary, while Grew was voted into Henshaw's place as the other Secretary; with both Curators now Secretaries, the office of Curator lapsed for a few years. Brouncker was voted out of office as President to Hooke's delight; Hooke had canvassed busily beforehand for Wren, not to be elected President until 1680, but was content with Sir Joseph Williamson, Secretary of State (and long Oldenburg's patron), although he had no real interest in natural philosophy or the Society's affairs.

With Oldenburg dead and Hooke Secretary, things changed more from Hooke's point of view than from the Society's. Hooke was now rather readier to perform experiments than he had been for the past couple of years; noticeable too is that the minutes record all Hooke's experiments and remarks more fully than had been the case for many years, during which Hooke had failed to supply the necessary material. The greatest change in the meetings arose from the formal reading of the minutes of the previous meeting at the commencement of the meeting; if this had been done before it was not recorded. This provoked much discussion but might, naturally, produce an extreme of discursiveness.

At the first meeting of the autumn (already mentioned) Hooke, when asked, volunteered to show an experiment with a hydrostatic balance, described at length in the minutes together with the comments of those present. He also described a French invention for water-proofing leather and 'shewed a Portugal onion which he had received from Dr

Presidents

1677–80	Sir Joseph Williamson
1680–2	Sir Christopher Wren
1682–3	Sir John Hoskins
1683–4	Sir Cyril Wyche
1684–6	Samuel Pepys
1686–9	John, Earl of Carbery (Lord Vaughan)
1689–90	Thomas, Earl of Pembroke
1690–5	Sir Robert Southwell
1695–8	Charles Montague (later Earl of Halifax)
1698–1703	John, Lord Somers

(1)	Secretaries	(2)	
1677–9	Nehemiah Grew	1677–82	Robert Hooke
1679–81	Thomas Gale	1682–4	Robert Plot
1681–5	Francis Aston	1684–5	William Musgrave
1685–7	Sir John Hoskins	1685	Tancred Robinson
1687–1709	Richard Waller	1685–93	Thomas Gale
		1693–1713	Hans Sloane

Note: Aston and Robinson, elected 30 November 1685 as customary, resigned, necessitating new elections, 16 December 1685

Clerks

1662–84	Michael Wicks
1686–99	Edmond Halley

Curators of Experiment

1682–1703	Robert Hooke
1683	F. Slare and E. Tyson
1684–7	Denis Papin

Operator: Henry Hunt

Editors of the *Philosophical Transactions*

1677–9 Grew	1685 Musgrave	1692–4 Waller
1683–4 Plot	1686–92 Halley (as Clerk)	1695–1713 Sloane

Figure 5. Officers, 1677–1703.

Whistler',[4] after which Philip Packer[5] (F.R.S. 1661) described a newly patented cider press which he claimed to have registered with the Society 'some years before', and which provoked 'much discourse concerning cider' as in earlier years. Hooke then read a letter from Leeuwenhoek, uncertain what he should do now that Oldenburg was dead; consideration was postponed to the next meeting but Hooke could not resist decrying Leeuwenhoek's microscopes and being asked to make one 'very likely to do as much', since he said he could. Finally, Hooke 'produced' an ephemeris of the eclipses of Saturn by the moon sent to him from Hamburg. It has seemed worth considering these minutes in detail, to show clearly that, except that Hooke's contributions are minuted more fully than those of anyone else and the minutes give the impression of being undigested, this meeting was itself in no way different from those of earlier years. And this is even truer of the minutes of the meetings on 1 and 8 November, where the record is very full, with far more detail than had been Oldenburg's custom to give, especially in the details of Hooke's experiments (one to test Leeuwenhoek's observations, one showing his own version of waterproofed leather, and one with his hydrostatic balance), and in reporting the minutiae of discussion. In contrast to these meetings, the minutes for the meeting of 15 November consist of only one paragraph, an incomplete account of Hooke's repetition of Leeuwenhoek's discovery of 'little animals' in pepperwater.

After the Anniversary Meeting and the elections of 1677, the Society settled down in more regular fashion. Hooke and Grew divided the secretarial duties between them: Grew seems usually to have taken the minutes, while, as his *Diary* often records, Hooke wrote them up later.[6] But as he noted (13 December 1677) 'It seemed as if they would have me still curator, Grew secretary'. And although Hooke remained Secretary, the Society did indeed wish for the continuation of experiments: at a Council meeting on 19 December

> It was ordered, that what experiments shall be undertaken by the Curators shall be propounded a fortnight before the showing thereof, that objections, answers, and confirmations may be timely thought of; and
>
> That the Curators or any other person shewing an experiment to the Society shall explain the same, and shew the usefulness of it.

It is difficult to know quite what to make of these commands, except to suggest that it was proving hard to preserve continuity in the meetings, discussion becoming all too random. Perhaps too the calibre of the Fellows was lower than it had been, so that they did not always understand the point of at least some of the experiments presented. Certainly many of the then active Fellows – like Whistler or Packer or Oliver Hill – are now little remembered, not having been active earlier, while the presiding officers were generally not so capable as Brouncker had been of taking the lead in discussions. But in fact, adherence to the proposed formalities was never strictly enforced and discursive discussion continued to be the norm.

Generally, Hooke's activities in 1678 fill a disproportionate amount of the meetings compared with previous years, if the minutes are to be trusted as a balanced record. Neither he nor Grew gave any lectures or discourses and there were few long papers or letters by others, exceptions being a speculative paper by Oliver Hill (see below, Chapter 9), which gave rise to much discussion, a long letter from Leeuwenhoek which was read in parts for several weeks,[7] and an account of hurricanes brought in by Southwell but composed by a Captain Long. It was clear that steps were needed to ensure that more letters were received, and it was resolved that when received they should be speedily read.[8]

The new year, 1677/8, opened with more discussion than concrete fact. Thus on 3 January the minutes of the previous meeting produced a long and speculative discussion about air, with Hooke propounding an aetherial explanation for its properties, especially that of change of barometric pressure, while Hill read the extremely speculative paper already referred to; otherwise this meeting was chiefly taken up with reading 'the epitome of six papers from Mr Hevelius to the Secretary' (that is, to Oldenburg, of whose death he had not yet heard), the letters being of course mainly devoted to astronomical observation, and with a letter from Swammerdam (also to Oldenburg) about viviparous snails. Neither produced any recorded discussion. In contrast, the discussion about air continued vigorously at the next meeting, when Wren

> doubted, whether there were any such thing as that aether, which Mr Hooke had hypothetically supposed; and said, that he would gladly see some experiment, that would make it evident, that there is such a body mixed with the air.

Characteristically, Hooke also professed to be in favour of such experiments, but saw no need for any fresh ones since, he claimed, 'he could by hundreds of experiments evidence the reality of such a body', whose properties were thereby revealed. He spoke of having a list of experiments

> which he thought he should have occasion shortly to make in order to the elucidating a theory, which he designed to make public hereafter.

Clearly when Hooke here said he could give evidence of experiments which proved the existence of the aether, he meant he could conceive of experiments which would do so, not that he had as yet *performed* 'hundreds' of them. It is worth remarking that no more than this exchange is given in the minutes (not always complete at this time), although at the next meeting it became plain that there had been read an account of animal parasites (worms found 'in the head and brains of some creatures'), an account by Plot of smut in wheat, an unsuccessful experiment on air (probably by Hill), and mention of a letter from Leeuwenhoek, while Hooke's *Diary* refers to his performance of an experiment 'of gravitation' and of 'a microscope'. At the next meeting (17 January) Hooke certainly performed what his *Diary* calls 'the Hydrostaticall experiment' and the minutes call

> his experiment, in order to explain the pressure of the air upon the mercury in the barometer,

this being followed by what the *Diary* calls 'much discourse' which the minutes recount in detail, including Hooke's theory of the air's elasticity. (This was actually less a theory than a statement that the greater the pressure, the more the air is condensed and hence the stronger its spring, a statement which had been experimentally verified more than fifteen years previously by himself, Towneley and Boyle.) Hooke seems to have been pleased by the ingenuity of this particular version, for he repeated it with variations at later meetings.

Generally during 1678 there was rather more than for some time of experiment and demonstration (including microscopy) most of the experiments being performed by Hooke, although others reported on their private experiments. But at the same time there was a great deal of general discussion about causes of phenomena revealed by experiment and theories to explain them than had been the case in previous years (at least as revealed by the minutes). The minutes for this year are a little discursive, but otherwise much as they had been five years earlier, with the presentation of experiments, comments and the reading of papers and letters. The chief differences are, as in the previous year, that the reading of the previous week's minutes and any discussion arising from this is always recorded, Hooke now supplied details of his own activities and inventions, and his theories are written out at some length.

In May 1678 a single experiment replaced two meetings. This is curious in itself, and the more curious because the experiment was so well known and tried. It consisted in observing the length of a column of mercury at the bottom and top of a considerable height, in this case the

200-foot 'column' or 'pillar' on Fish Street Hill, the well-known Monument usually thought to have been designed by Hooke, and finished in 1676. Hooke's *Diary* shows him to have tried the experiment himself on 16 May and to have 'directed' Hunt in performing it a week later. As he told the Society's meeting on 30 May, when he read an account, he was not satisfied with the accuracy with which the intervening heights were measured; it was intended that the experiment should be repeated, but when he began to prepare for it, he 'Brake the mercury glasse in the Parlour' and the plan was dropped. This was not his only mishap at the time, for apparently at the next meeting there were too few present to make minutes worthwhile and he was soon to record in his *Diary* great anger at both Boyle's sister, Lady Ranelagh, and George Ent (F.R.S. 1677) which no doubt distracted him.

In later June, however, meetings revived, and although no experiments were performed (although some were 'appointed') the discussions and presentations were reasonably supplied with references to experiments and observations made previously. In July, Hunt was ordered to try the experiment of plants growing in water without air, which Croone thought possible and Hooke denied; Hunt's trial duly showed that Hooke was correct. This led, via a discussion of the necessity of air for various creatures, to the physiology of cold-blooded animals like eels, from which the discussion wandered to whales, to strange herbal decoctions, and their use as antidotes, which led to discussion about poisons. The train of discussion is easy to follow so far, but then Cox, apparently suddenly, related an experiment made by himself on salts from snow water, a clear example of the unregulated discussion which occurred all too frequently. Poisons were always a popular subject and were a topic for discussion at several subsequent meetings. In August Hooke read his Cutlerian Lecture on springs and showed a few confirmatory experiments, after which this long session closed, the meetings being resumed only at the very end of October.

The autumn meetings were mostly filled with discussion arising from the reading of letters, as it happened mainly concerned with natural history, always welcome. There were a few experiments: on 14 November an attempt was made to see if a cup of ivy wood would separate a mixture of wine and water into its component parts, a failed experiment since the cup proved to be impermeable; Hooke showed 'his experiments which were divers ways of making very round and clear globules of glass for microscopes with great ease' (i.e. Leeuwenhoek microscopes); and after the Anniversary Meeting he showed a series of demonstrations with the condensing engine (which had been out of action but was now repaired); these included comparisons of the effects of condensed and rarefied air on birds, a much-tried subject by this date.

Something over a year since the change in the Society's administration in which Hooke had been the leading spirit, he himself saw the situation as regards experiment as unsatisfactory. It has often been suggested by historians that he was overburdened by the orders and demands of the Council, but it seems more likely that, on the contrary, he had revelled in his rôle as Curator of Experiments and was far happier in that office than in that of Secretary, which he was to relinquish in 1682. It is true that he often found the assembling of the necessary apparatus an unwelcome burden, but that was the reason for wanting an assistant. Henry Hunt, hired in 1673, was now growing ever more competent and could usually be trusted to take this task over, or even to perform experiments which Hooke and others devised. Hooke evidently could still always think of experiments to illustrate any topic raised and certainly believed that more experiments should be performed, both at meetings and in his own rooms. Consider the minutes of the meeting of 7 January 1678/9, which record that

> Mr Hooke gave an account of the experiments which had been shewn at the last meeting, and the design of them, and of the time and manner requisite for the completing such experiments; and he desired, that there might be a committee for the making such trials, as could not be made within the time of the sitting of the Society, since many experiments could not be made within so short a time.

Hooke's desire illustrates some of the problems associated with the public showing of experiments, including lack of time and lack of resources. At the next meeting, when he gave 'an account of the design of an experiment about respiration', he tried to get a committee formed to perform this experiment, but although there was much discussion and a committee was indeed formed, nothing more is heard of it. Undaunted, Hooke was suddenly very active, showing experiments at five meetings in January and February to illustrate his 'theory about fire', experiments which pleased him as his *Diary* records, and which provided entertainment and discussion at the meetings.

Perhaps stimulated by Hooke, the Society at this time canvassed a number of subjects suitable for discussion and experiment. At the very end of February 1678/9 it was noted that

> The experiment shewn by Mr Hooke was to shew, that vapours press only according to their own gravity, and not according to the space, which they take up in the atmosphere.

(Unfortunately, this was less than wholly successful, and not everyone agreed that it showed what Hooke said that it did, and although it was decided that it should be tried again, Hunt was 'not able to procure convenient glasses'.) Combustion was an easier subject to deal with and continued to provide experiments (mostly familiar) and discussion. Then it became plain[9] that not everyone agreed with Hooke's 'hypothesis of explaining it by the dissolution of bodies in air', but it had provided experiments during two months, and so was useful. The minutes for March are very sketchy, and do not include the simple experiment Hooke recorded in his *Diary* for 27 March. The attention of the members was now directed into other paths by the mention of a reference to gunpowder found in the works of Roger Bacon, a name which, then as now, commanded equal parts of attention, interest and doubt when connected with the introduction of gunpowder. Experiments in April tended to be simple: on 3 April 'Mr Slare presented the Society with a phosphorus of his own making', which was 'examined by several present, and judged to be as good as any, which they had seen' but no details are given. Henshaw, confirmed by Hooke, reported on its action at several meetings but said nothing of its composition. During April and May Grew communicated anatomical observations and at many meetings read sections of his 'description of the repository' (to be published as *Musaeum Regalis Societatis* in 1681), hardly experimental. Thomas Allen, F.R.S. 1668, gave an account of mineral water analysis, and Hooke showed 'an experiment in mechanics' consisting of the exhibition of a rain gauge and a wooden hygroscope, while (8 May), when Croone cited an item from the 'Paris Gazette' (the *Journal des Sçavans*) about a reportedly successful attempt to fly from the top of a church steeple to the ground, Hooke produced 'a model of the contrivance of the wings [of birds] made with pastboard', which demonstrated the mechanism of flying and which in his *Diary* he called 'Experiment of French Winges'. Here, as often, Hooke uses the word 'experiment' in, presumably, the older meaning of trial or test; certainly he and so presumably others thought that the showing of models and of instruments, especially when a demonstration and/or an explanation of their uses was involved, counted as experiments. (In the same way it seems clear that the showing of an anatomical observation counted as an experiment.) This attitude must be borne in mind constantly when assessing the state of experimental demonstration at any particular period of the Society's existence.

At the end of May 1679 there appeared on the scene one who was later to serve the Society as an active Curator of Experiments. This was Denis Papin, who had been assistant to Boyle since 1675, in which capacity he was known to some Fellows, while his work earlier with Huygens had been discussed at a meeting of the Society in 1676. He had invented a new style of airpump with a water seal which he had used for the experiments which Boyle had devised. Now he also had a new invention in his 'digester', a pressure cooker which was found capable of softening bones and reducing many things to a state of surprising tenderness. Hooke was responsible for asking permission for him to come before the meeting of 22 May to 'shew an experiment to the Society'. This was less an active experiment than the result of an experiment, namely a sample of hartshorn softened in his digester; it aroused so much interest that Papin was to be requested to attend most of the meetings in June and July to display more examples of digested materials, as also (19 June) 'a new kind of wind-fountain of his own contrivance'. Hooke saw a great deal of Papin during the course of the summer and himself utilised Papin's airpump for an experiment with a spring, which he found to be unaffected by air pressure, which encouraged him to believe that watch springs would also be unaffected by changes in barometric pressure.

Hooke had already (3 July) suggested that the Council should employ Papin as an amanuensis.[10] This was done, although he was paid only when (22 September) it was reported (by Hooke) that he was about to leave for Paris. He was then offered twenty pounds a year and free lodging in Gresham College.[11] He appeared twice before the Society, bringing his windgun and showing Boyle's discovery that fruit could be preserved *in vacuo*, but on 10 December the Council decided to discharge him. Nevertheless he continued to show experiments occasionally during 1680, maintaining contact with the Society, presenting his book (*A New Digester*) in 1681 and being elected a Fellow in 1682.

The Council expressed itself once again as dissatisfied with the content of the meetings. On 8 December 1679 it was resolved to try to limit discussion to one subject and when 'they are satisfied concerning that subject' it should be written up for publication. Besides this, at each meeting an experiment should be 'appointed' to be shown at the next, so that those present could be prepared to discuss it. After each meeting,

the Curator should prepare a careful account of the experiment performed, to be read at the next meeting and entered in the Register Book
(which had apparently been neglected). And the Council was to check
monthly that these regulations were being complied with. Alas, like
most such resolutions, this was never pursued, and indeed 1680 was to
prove a low point for Hooke as far as experiment was concerned, even
with urging in the new year by Petty and Wren, which was presumably
the reason for his recording in his *Diary* 'much cavilling and cabaling'
and 'Croon, Gale, President, Colwall, Henshaw, against me'.[12] But he
was stimulated to more activity in the coming months, showing a series
of measurements of the specific gravity of metals, accurately determined, which he linked with the structure of matter; his experiments
were apparently partly performed at meetings, partly privately, but all
provided material for discussion. At the end of May he showed an
experiment 'found out by his highness Prince Rupert' in which a very
hot flame was created by heating brandy in an aeolipile, novel only for
its use of brandy; Hooke fairly claimed that he had used an aeolipile long
before 'for driving out the air of small round glasses'. (Such devices had
long been used to create a forced draft.) Henshaw (8 July) introduced a
number of experiments on the motion created in a glass of water whose
rim was rubbed with a wet finger, as described in Bacon's *Sylva Sylvarum*,
while in December the meeting examined 'the electricity of glass after
Mr Newton's method, by rubbing one side of a glass to make the other
attract'. During the year several experiments with capillary tubes were
tried while, as noted, Papin showed his digester and some few other
experiments, and a letter from Leeuwenhoek was read at length. But it
was on the whole a poor year for experimentation.

1681 was also to prove a poor year for experiments, whether performed at the Society or described as performed elsewhere, and as the
minutes are longer than in 1680, this must in fact have been the case. At
the beginning of the year there was some interest in magnetism and (19
January) Hooke produced an instrument

> for making experiments in order to find out the attractive power of the
> loadstone at several distances, and to reduce that power to a certain
> theory

but there is no record that he made the proposed experiments. A
fortnight later a paper by Haak was read, giving an account of his
experiments 'about recovering and increasing the attractive virtue of the

magnet'. Several groups of experiments were pursued in the spring. The first turned on Papin's digester which he had given to the Society when he left England; it took some time for (presumably) Hunt as Operator to learn to use it properly, but once he had done so various substances were softened in it for exhibition. The second group of experiments derived from Southwell's presentation of a sample of phosphorus (the modern element), source unstated; curiously, no mention seems then to have been made of Boyle's sophisticated chemical explorations of this novel material in his two related works *Aerial Noctiluca* (1680) and *Icy Noctiluca* (1681/2), although the experiment 'sent by Boyle' and performed on 7 December of producing light by a mixture of two fluids was derived from his work on phosphorus which he had been the first in England to isolate. A third group of experiments concerned air resistance as measured by a pendulum and was accurate but not novel. And in July Hooke showed how to make sounds 'by the help of teeth of brass-wheels', and showed two instruments. Several meetings were further taken up by Hooke's long discourse 'about the nature of light and luminous bodies', a continuation of lectures read the previous year, an interesting essay but non-experimental.[13]

In March of the next year, 1682, Hooke showed a simple experiment intended to demonstrate that the heat of a fire was propagated differently from the heat of the sun, by showing that a plain glass plate placed between 'the concave metal' and the fire kept the heat off the focus of the mirror, but not the light. This ingenious demonstration was novel, but hardly the conclusion that heat and light, although essentially the same, are propagated differently. Hooke also showed several optical experiments more amusing than instructive.[14] In May his discourse on light was continued with a discourse on its method of propagation, which led to another on local motion, other discourses which he read in 1682 being 'concerning the means, how the soul becomes sensible of time', a long essay which attempted to explain the 'organ of memory', and a discourse on the divisibility of matter, supported by an experiment which he performed on 20 December which purported 'to shew the easy divisibility of the body of the aether', in fact very similar to an experiment described in print by Boyle in 1669,[15] which had been intended to show that light bodies fall much more quickly *in vacuo* than in air but now applied to 'explain' the nature of the aether (which Hooke took for granted) and the motion of comets' tails. In the autumn Hooke read several times on the several comets of 1680 and 1681. Otherwise, experimental content came mainly from the reading of Leeuwenhoek's letters and from a few scattered experiments shown at meetings. Thus in late February Slare showed some experiments on phosphorus; in the late spring and early summer several meetings saw the repetition of experiments on strength of materials, in which boards made of different kinds of wood were broken with carefully measured weights; and at the end of October Hubin, a Paris instrument maker (said in fact to be English and certainly known to Hooke) shown three *soi-disant* experiments, really demonstrations of three hydrostatical toys, including an Archimedean diver.[16] During the year the medical men continued to read papers on anatomy and physiology, more or less experimental, and on natural history. But on the whole, 1682 like 1681 was a poor year for experiment as such, although discussion in which experiment was cited was often brisk.

St Andrew's Day, 1682, however, saw a number of changes in the Society's administration which were to have a very considerable effect upon the experimental content of meetings. It is difficult to understand exactly how the changes came about, but certainly a new turn was given to affairs by the replacement of Hooke as Secretary by Robert Plot and at the same time by the activities of Aston who had replaced Grew in the previous year. Aston now revived the *Philosophical Transactions*, as Hooke had refused to do, while Plot, although resident in Oxford, kept in constant touch with the Royal Society at the same time that he was extremely active in the Oxford Philosophical Society. This, which had first been formed about 1651, seems in the 1680s to have had a new lease of life, virtually a re-foundation, and to have functioned in much the same way as the very early Royal Society and been indefatigable in the performance and presentation of experiment. Plot now contributed a great deal towards experimental discussion in the Royal Society's meetings by sending long reports of what was being done in Oxford. This however was not enough to satisfy all the Fellows, and on 10 January 1682/3 there was a familiar sounding resolution about regulating the preparation and writing up of experiments; it is referred to as a revival of Council orders 'made in the presidentship of Sir Joseph Williamson', that is, at the end of 1679. Perhaps it was hoped that, with more leisure, Hooke would revert to being really active as Curator once again. For the next few meetings Hooke did indeed bring in experiments. But as he soon ceased to do so the Council evidently felt it necessary to force his hand by direct action. On 6 June the Council resolved that Mr Hooke shall receive every meeting day order for the bringing in two experiments at the next meeting day, together with a declaration by word of mouth of the purpose and design of the experiments and an account in writing of the history thereof, and the purpose as aforesaid, as may be fit to be entered in the register: and at the end of every quarter there shall be a meeting of the Council, where his performance shall be considered, and a gratuity ordered him accordingly; and that from this time he shall have no other salary.

All too evidently, newer Fellows mistrusted Hooke's willingness to perform experiments unless forced in some way. As he was not present at the Council meeting, being no longer a member of Council (nor apparently at the ordinary meeting on the same day), and his *Diary* for this year is not extant, his immediate reaction cannot be known. But he soon gave over any anger he may have felt (and most likely did feel), for on 20 June he 'declared his satisfaction ... , and his resolution to proceed in the office of Curator'. And in the event he was far more active in 1683 than at any time since 1679.

The Council went further. At its meeting of 28 February 1682/3

It being mentioned, that the Society wanted experiments at their ordinary meetings, Dr Tyson and Dr Slare were proposed as persons very fit to assist the Society in that work,

Tyson to make 'anatomical dissections and observations', Slare 'chemical and other experiments'. (Tyson had been a Fellow – proposed by Hooke – since December 1679 and had been elected to the Council in November 1682 and was to assist Plot in work on the *Philosophical Transactions*. He was an Oxford M.D. and was already well-known for his work on comparative anatomy. Slare had been elected a year later than Tyson and was not on the Council; he received the M.D. degree from Utrecht in 1679 and from Oxford in 1680. In 1679, when he presented a sample of his phosphorus to the Society, he was associated with Boyle in chemical work; later he said that he had learned all he knew about chemistry from Boyle, with whom he must have been still working in 1685 when he communicated some of Boyle's work to the Society.) The records show that both Tyson and Slare were active in presenting material to meetings, if not quite so active as to present something to each meeting as hoped. But Slare showed more experiments than Hooke did in this year. However, by far the greatest experimental contribution came from Oxford, whence came a stream of letters reporting experiments performed there, ranging widely over many fields: physiology, mineralogy, mechanics, pneumatics, chemistry, hydrostatics. Each report produced discussion at the London meetings and occasional emulation and repetition. Independently, a good deal of attention was in this year paid by the Royal Society to magnetism. Thus in April 1683 Hooke performed 'an experiment tending to explain magnetism' – but the minutes do not explain the explanation – while a fortnight later 'an experiment formerly shewn the Society by Mr Henshaw was again exhibited': it related to magnetic polarity and was more fully described at the next meeting. Then on 23 May

> Mr Halley having a new hypothesis concerning the variation of the compass, and the magnetism of the earth, made some experiments for the better explaining it,

experiments which are described in the minutes in some detail. (There was an accompanying paper of theory which was printed in the *Philosophical Transactions*.)[17] In the autumn (late October and November) Hooke showed several devices: his anemometer, an instrument for measuring the rate of flow of water, and also 'a convenient way of copying any thing'. He also promised to write up the experiments he had shown during the year with 'an idea of natural philosophy built upon them', but which he said he could not do without money to assist in the repetition of the experiments. The Council approved his design and promised him money; a fortnight later (12 December) he was asked to supply an account of 'the experiments made by him as curator during the last half year', which, when he had provided it, led to payment. This year, 1683, shows more experiments performed, both by Hooke and by others, than for many years; Hooke indeed performed experiments at approximately one-third of the meetings and described the same number; Slare was equally active; while the accounts from Oxford greatly increased the number described at meetings.

The balance of experimental performance, however, was to alter radically in the years to come, when Hooke's interest shifted to the reading of papers and the description of instruments. Others were to replace him as principal Curator, but none was so assiduous as he had been at the peak of his endeavour. To balance this in some measure, for a few years the Dublin Philosophical Society began to rival Oxford in providing long accounts of experiments performed by them, accounts for the Royal Society to discuss and comment upon. Many of Hooke's accounts of experiments performed by him in private but reported upon at meetings are preserved in the Register Book for 1683/4, in response to the Council's demand (23 January 1683/4), but the number of those which he performed at meetings steadily declined. Consequently, at the end of February the Council asked Halley 'to bring in experiments at meetings of the Society in the manner of a Curator', told him that he might be considered for the post, and requested that he should continue his series of experiments on magnetism. The immediate result was not that Halley performed more experiments but that Hooke did, and these were mostly also on magnetism. But these were not really very numerous, and to make matters worse, Tyson and Slare were never again to be so active as in 1683, although Slare particularly remained a steady contributor for some years.

The Council was understandably concerned, and when, at the beginning of April, Papin offered himself as temporary Curator for the remainder of the year, the Council promptly accepted the offer and paid off Hooke with a quarter year's salary. Papin was to prove an energetic although somewhat limited Curator, devising many acceptable and often very ingenious experiments with his two instruments, airpump and digester. (For example, in April he showed how he could make plaster and other substances transparent by making turpentine penetrate their surfaces.) During the remainder of 1684 Papin performed at most meetings, either showing experiments and/or reading experimental papers or discourses, all thoroughly within the terms of his employment. The meetings were enlivened further by various Fellows. Lister presented a paper on meteorology, full of observations, and performed a number of simple experiments, including the demonstration of the crystal structures of salt and of ice, while under his direction the Operator (Hunt) performed experiments on magnetism to illustrate a paper presented by Lister (27 February), experiments which Hooke later repeated. (Otherwise Hooke's attention, released from devising entertainment for the meetings, turned to such matters as Roman archaeology, on which he read a paper in July.) Tyson showed the anatomy of a viviparous snail from the Thames (4 June). Slare reported on experiments made on 'sea ice' during the winter together with several freezing experiments, and in November he repeated three chemical experiments derived from Kunckel's *De acido & urinoso Sale* which was dedicated to the Society.[18] On the whole,

Papin was the most active presenter of experiments, although noticeably inactive in discussion if the minutes are to be trusted.

These changes continued in 1685 when Papin's appointment was renewed. During the whole of this year Hooke performed no experiments at meetings nor did he read any papers or lectures, even though he attended meetings regularly, made comments, reported past observations and (17 June) 'shewed an instrument for the drawing the logarithm line' adding characteristically that Descartes had 'supposed [it] not to be practicable'. Papin was extremely diligent throughout the year, producing one or more experiments at virtually every meeting and also reading a number of papers, each of which contained accounts of experiments which he had performed elsewhere. Slare was perhaps desirous of emulation, hoping for a similar appointment, for (22 April) he proposed to the Council

> a design for a chemical operator to the Society, who should attend at their meetings, and be content with a moderate salary,

a motion which was approved but never implemented. Not until the autumn did Slare himself take an active part at meetings, producing first (28 October) a paper by Boyle about a 'strange self-moving liquor' later published in the *Philosophical Transactions* (the fluid had not been discovered by Boyle but he did make some observations of it under different conditions), and then, a month later, a substantial paper of his own on 'the insufficiency of alcalis and acids to discriminate the *res medica*', and finally, in December, showing crystals of amber.

After this burst of activity in 1685, Slare became inactive in 1686. But Papin was only marginally less energetic than during the previous year, showing fewer experiments but reading many papers concerned with experiments which he had made in his digester, in the airpump and in condensed air in a windgun, so much so that the experimental content of the meetings was really dominated by his work. Hooke contributed half a dozen accounts of instruments and ways of improving them,[19] while in December he turned his attention to fossils, producing a very long paper illustrated by 'very elegant figures of those substances drawn by himself'. He had already (in the spring) produced 'an analysis of the whole matter of hydrography', which, unlike the paper on fossils, could not be counted as really empirical. In spite of being urged to do so by the Council (3 March), he performed no public experiments in this year. Generally the meetings contained much discussion arising from papers and letters read to them, on subjects ranging from the tides to animal anatomy, astronomy and experiments reported from Oxford and Dublin, but few experiments were presented by the Fellows. Halley, now Clerk (a post which forced him to resign his Fellowship for its duration) read 'an account of an experiment made by himself to find the comparative weight of quicksilver to water by weighing a quart of quicksilver in water (14 April), whereupon Hooke 'shewed an experiment for finding the same thing by a syphon filled with mercury in one shank and water in the other', producing a value nearer to the true one than Halley's. (This was the only experiment which Hooke showed in 1686.) Halley also read a discourse 'concerning the cause and properties of gravity',[20] a serious work, but more theoretical and mathematical than experimental.

At the beginning of 1687 (5 January), the Council once again tried to get Hooke to commit himself, to declare what his intentions were in regard to his post as Curator. In reply he proposed that he should provide two experiments and a discourse at every meeting 'provided his salary be made up [to] 100 l. per ann.', as the minutes put it. The Council offered fifty pounds, promising that they would help him get the remaining fifty from Sir John Cutler for Cutlerian Lectures, at the same time asking him to ensure that 'the said experiments ... proceed in a natural method'.[21] Hooke did produce a goodly number of discourses at meetings, that is at about half the meetings (although by no means all were on different subjects): they ranged from considerations about the shape of the earth (with his ideas for a telescopic determination of the true meridian line) to a long treatise on fossils and conclusions about their distribution, how to measure exactly an interval of time (by dividing the time of vibration of a pendulum 'into its parts') to a long discussion of classical geography in the shape of a commentary on the Greek periplus attributed to the Carthaginian Hanno. These all aroused great interest, but although some contained empirical evidence none were experimental, and indeed Hooke performed only a couple of real experiments during the whole of this year, suggesting a genuine decline in interest on his part. The experimental content of the meetings, which was not inconsiderable, was chiefly provided by Papin, who either performed at least one experiment or read accounts of experiments performed by him at most meetings until mid-November, when he left England for a post at Marburg. Halley was active during the earlier part of 1687, once again discussing magnetism and offering an experiment 'for finding the comparative force of a loadstone at several distances'. Discussion arose both from the reading of the minutes of previous meetings and from letters, which, in the summer, included accounts of proceedings in Dublin. At the beginning of the year (January) Wallis sent a letter about the resistance of a medium to projectiles; this led to a curious suggestion that Newton be asked 'whether he designed to treat the opposition of the medium to bodies moving through it in his treatise *De Motu Corporum* then in the press' – curious, since obviously Halley knew all about the contents of Newton's *Principia* which he was seeing through the press. Wallis's interest had presumably been aroused by the preliminary version sent to the Society at the end of 1686, while, when Wallis wrote again on the subject (9 March), 'a paragraph of Mr Newton's mathematical philosophy' relevant to Wallis's letter (commenting on Hooke's 'hypothesis of the mutability of the poles of the earth') was read to the meeting.[22] To a certain extent, both the Society and experiment languished in 1687: on three occasions there was no meeting because no presiding officer was present, while once there was a Council but no ordinary meeting.

Papin's departure meant a decline in the showing of experiments not to be arrested until the end of 1703, when Newton's election as President brought the appointment of a new Curator of Experiments, Francis Hauksbee. Not that the Fellows lost interest in experimental philosophy during these fifteen years, but rather that the Society lacked leadership: the Presidents, of whom there were five between Pepys (1684–6) and Newton served short terms and were not primarily natural philosophers,[23] the Secretaries Waller (1687–1709) and Gale (1685–93) were literary, not philosophical, although Waller was keenly interested in promoting works by others and helped to revive the *Philosophical Transactions*, while Sloane, the other Secretary between 1693 and 1713, seems not to have played an active rôle at meetings until after 1700, being junior to most of the Fellows in status and years at this time. Hooke was aging and now far less interested in the labour of pure experiment than once he had been, and in spite of his repeated agreement to provide experimental entertainment for the meetings he was inclined to be satisfied with reading papers which were more theoretical than experimental, although of course based upon a lifetime's familiarity with experiment. Indeed, his papers were often decidedly speculative and frequently began as accounts of books to which he added his own views on the subjects upon which they touched.

Further, as his later *Diary* (1688–93)[24] reveals, Hooke was often out of sympathy with other Fellows who produced papers, disliking Halley, Dr Havers (F.R.S. 1686, known for his work on bone structure), Slare, the French instrument maker Grillet, and Newton's protégé Fatio de Duillier almost equally.

In 1688 a good deal of the experimental content was provided by Halley.[25] Although, as noted, he was officially only Clerk and never Curator, he played a major rôle at meetings, as he was to continue to do throughout his tenure of office. Hooke's list of members present at meetings nearly always included Halley's name,[26] and the minutes record him as active at fifteen of the meetings in 1688, more in subsequent years. He then read ten papers on a variety of subjects: mathematics, astronomy, meteorology, 'an attempt to explicate the rising of vapours out of water', longitude determination, the equation of time, metrology, and instruments. He showed the Society only one experiment in this year, which demonstrated by combustion that cochineal was of animal composition, but he also produced an account of some experiments 'to ascertain the quantity of the dilation of fluids by heat' made by himself. He further (February) reported on a French treatise on 'Barometers, Thermometers and Hygroscopes', noting the account therein of the flash of light produced in the vacuum above the agitated mercury in a barometer when it was carried (discovered originally by Picard in 1676 and then described in the *Journal des Sçavans*); on this occasion Hill correctly related it to the phenomenon described by Boyle when the key controlling the valve between the receiver and the evacuated cylinder of his airpump was opened.[27] Besides Halley's work, only five experiments were shown in 1688, of which two were performed by Allen Moulin (or Mullen) of the Dublin Philosophical Society, who, in a singular burst of activity, read three papers which he illustrated; one further experiment was performed by 'Mr Boyle's man'. There were several experimental discourses besides those noted above, including Tyson's account of anatomical discoveries and several letters from Leeuwenhoek.

Early in 1689 Hooke gave a long series of lectures on geology, having been asked by the President for an account of Burnet's *Sacred Theory of the Earth*; these lectures aroused interest and discussion but provided no occasion for experiment. But on the whole, experiments were more numerous than in the previous year, being performed at a dozen meetings while another dozen at least were described. Hooke himself performed two (unusually they were quasi-chemical, involving a mixture of vitriol and water in an attempt to understand what he called 'the Penetration of Bodies', from which he argued that more than mere mixture was involved), reported on three more and, besides his geological lectures, discussed (10 and 24 July) the value of experiment in advancing natural philosophy, and mentioned various potential inventions of his own. Halley was a little more active, showing experiments on three occasions and reporting his experimental investigations at five meetings, besides reading three theoretical papers. Halley's experiments, which arose out of discussion, mainly involved a comparison of sea salt and sal gemmae (rock salt); he studied their solubility and the specific gravity of their solutions, concluding that they were identical. He also watched, with the aid of a microscope, salt crystallising out of solution, reporting the shape to be that of a square and all the 'atoms' (i.e. crystals) to be of equal size. Waller, Henshaw, Slare, Havers, Moulin and a Mr Watts all contributed at least one experiment, performed or reported, while Papin sent a letter which aroused interest. On the whole, the experimental content was thus relatively high.

But this was a flurry of activity not well maintained in the succeeding decade. More and more in the 1690s the substance of the meetings was provided by discourses, Hooke's among them; letters; the production of natural history curiosities; and discussion of the previous week's minutes, all very much in the style of the previous decade but generally more discursive and less organised with less innovative content than in the past. Those whose names appear making contributions for the first time mostly made little impact, there being apparently no rising generation to provide notable Fellows.

The decade began badly. In 1690 Hooke made no experimental contributions whatsoever, his eight discourses being mainly theoretical. Halley's activities were also mainly theoretical. He read five discourses on subjects not susceptible of experimental illustration: on the date of Caesar's landing in Britain, on the equilibrium of the moisture of the sea and the 'power of the sun's heat', on meteorology, gunnery and tropical natural history, as well as 'geometrical propositions' relating to optics and the determining of longitude at sea, his only experiment (5 November) being on magnetism. Only four or five experiments were actually performed at meetings during this year, while on only four occasions were experiments even discussed (by the instrument maker Grillet, not a Fellow, by Slare and by Waller). Havers read a discourse on anatomy, which may be classed as experimental. Fatio de Duillier read a paper on gravity. Altogether, 1690 was a thin year for experimental discourse.

In the next year, 1691, experiments were still few, but less so than in 1690. Halley as usual was the most active in this regard. He showed experiments on refraction (involving a diamond and a piece of Iceland spar); he performed an experiment arising from his theoretical paper on the velocity of a bullet with reference to Newton's *Principia* (Book II, prop 37); and in the summer he performed trials with a diving bell of his own invention, reporting on them at several meetings, together with descriptions of instruments of potential use in diving, such as one for keeping a flame burning under water. He interested himself more generally in marine matters, describing pulleys for raising weights at sea, a device for blowing up the decks of ships and an anemometer, as well as discussing the physical cause of the rôle of air in the conveyance of sound, an hypothesis of the cause of changes in magnetic variation and 'a way of estimating the necessary swiftness of the wings of Birds to sustain their weight in the Air' – all very good value for his salary as Clerk, especially considering the fact that he also acted as one of the editors of the *Philosophical Transactions*. Hooke was in 1691 far less active than Halley, performing only two experiments (both on the rate of flow of water from a hole) and reading only four papers, mostly concerned with an instrument for sounding the depths of the sea (probably Halley's work was responsible for stimulating his interest in such matters). Slare was hardly active at all, reading only one paper (on chemistry, as usual).

The next year, 1692, was a better year as far as experiments and experimental papers were concerned: some dozen experiments were actually performed, and there were a number reported on as well. Boyle, although he had died in the previous year, provided some posthumous experimental interest when Sloane (although not yet Secretary) early in the year opened a sealed paper left at the Society's rooms some years previously: it concerned Boyle's discovery that a solution of silver in aqua fortis would reveal 'the least admixture of salt in water by rendring it milky', a fact which Sloane, apparently, confirmed by demonstrating it to the meeting.[28] Slare showed a chemical experiment also originating with Boyle, this from Boyle's work of thirty years earlier on coloured solutions; later he read a chemical paper of his own. Halley, as in the preceding year, was very active, speaking at no fewer than twenty-one meetings, but he performed no experiments, although once again he described a number of new instruments, navigational and astronomical. In March an experiment sent by an unnamed person 'from Reading' was described (by Halley): it concerned the rate of flow of water from a hole, which seemed to demonstrate that the fluid flowed faster from a small vessel than from a large one. It is not clear whether this had any connection with Hooke's account of the previous year, and no more apparently was heard of the unknown. John Houghton (F.R.S. 1680), known as an agricultural writer, showed the results of the quantitative distillation of beef in a retort, carefully examining the residue; his paper aroused sufficient interest to warrant publication in the *Philosophical Transactions*. Early in the year, Hooke read a discourse on the history of telescopes and microscopes, to be followed by several papers on miscellaneous topics and a number of book reviews. The scientific content of these reports was not high, although of course the subjects were of general interest to the learned world. It was chiefly Halley who maintained the content of physical science in 1692. At the same time biological science fared better. From Leeuwenhoek came a number of letters which were read on no fewer than nine occasions, all of which aroused interest and stimulated discussion. There were also numerous medical contributions, especially from Havers, most of high quality as regards empirical content. On the whole, however, the interests of active Fellows, even of Halley, were in subjects which did not lend themselves to experimental demonstration at meetings. Such subjects as astronomy, mathematics, meteorology and theoretical magnetism were all, usually, based firmly upon empiricism and observation but could not always be made to yield experiment for demonstration to meetings. The meetings usually displayed erudition and keen interest in natural philosophy, even of a quite technical nature, but the method of presentation had changed. There was also the obvious fact that the Royal Society, to all appearances, was broadening its interests in a way which, had it continued, would have made it very like a national academy of the Continental type but this was increasingly and perhaps necessarily at the expense of experimental learning.

This tendency continued throughout the 1690s, not always to the detriment of the scientific (as distinct from the experimental) content of the meetings. In 1693 four or five meetings were, as might have been expected, partly or wholly concerned with accounts of the eruption of Mt Aetna and the associated earthquake. The Society received numerous letters which were read at meetings (and many later published in the *Philosophical Transactions*); these readings provoked discussion and stimulated the production of several theories of the cause of earthquakes. Of these the most interesting are those by Flamsteed (Astronomer Royal) and by Hooke, the latter enshrined in a long paper entitled 'The Rape of Proserpina' in which he interpreted the Greek legend in terms of earthquakes, which he had already discussed at some length.[29] Earthquakes were certainly a legitimate interest for the Society, but they could not, obviously, lend themselves to experiment. Otherwise, there was the usual range of subjects. Leeuwenhoek's letters as always aroused interest, stimulating King to contribute a related paper on microscopy in general and the animalculae found by Leeuwenhoek in pepper-water in particular. Houghton showed a simple experiment on 'sensible plants', and there were a number of other papers on botany, as well as on natural history and agriculture. Boyle's directions for making phosphorus were read (and published); both Slare and Houghton showed simple chemical experiments; and Povey read a paper on painting with tempera. Halley continued active, reading papers on theoretical astronomy, on statistics, on hydrography, on nautical charts, and on 'Lobsters and Crabs shooting their Claws, and new ones growing again', with a reference to Boyle's having mentioned the same phenomenon. The experimental demonstrations were thus few in 1693, but it cannot be said that the Society's interests were either non-scientific or totally trivial.

As in previous years, in 1694 it was Halley who provided much of the Society's scientific interest, reading papers at ten meetings and speaking at more. These papers were partly concerned with mathematics and mathematical astronomy, but he also discussed the rate of evaporation of water under different conditions, the proper adjustment of a seconds pendulum, his 'explanation' of various features of the earth as arising from the impact of a comet 'or other great body', as well as his favourite topic, magnetism. Besides this, he mentioned unusual deposits of stones and oyster shells known to him, as well as giving an account of what Newton had told him about the increase in the earth's mass. Hooke was this year relatively inactive: he read only one paper, this on magnetism, and reported only once on instruments, although he did describe a couple of experiments which he had performed privately. Only twice this year were experiments performed at meetings, once by John Harwood (F.R.S. 1686) and once by someone unnamed, while Hooke,

Thomas Kirke, Cooper and Wallis described experiments which they had made elsewhere. Except for Halley, the number of those providing scientific papers was notably small.

The next year, 1695, was much more active once again. Halley described a curious windmill, read a couple of mathematical papers, a paper on the rate of flow of water from a hole (a recurrent topic), presented a calculation of the amount of water needed to cover the whole earth to a depth of two feet (a problem arising out of the Biblical account of the flood, a topic of Biblical geology much discussed by learned writers at this time), and an archaeological account of Palmyra. He was thus very active, with a typical admixture of subjects. Hooke's interest was this year mainly in instrumentation and microscopy, but he did perform one experiment at the Society's request, for the reading of Homberg's account of his preparation of phosphorus was followed by a demonstration of its properties by Hooke. At the end of May the Council decided to offer Hooke payment for every experiment he performed, but this failed to persuade him to undertake any more. Indeed, fewer than half a dozen experiments were performed at meetings this year: besides Hooke's, there was one by Harwood on a bituminous stone, an investigation by Hunt of sal gemmae, an anonymously performed trial of the effects of viper wine, and an anatomical demonstration by Tyson. Only five experiments were reported upon: these included a microscopical account by Hooke, Grew's analysis of spa water, an anonymous report on the rate of flow of water in pipes with different bores, and some clinical accounts by physicians. The greatest biological interest came once again from Leeuwenhoek, and there was the usual concern with natural history and agriculture. Parenthetically, strangers were now being introduced to meetings again.[30]

The level stayed much the same in 1696, with even fewer experiments performed at meetings (only two, one by Hooke, on the flow of water through pipes). There was, however, a considerable degree of interest in microscopes and microscopy, not only from Leeuwenhoek but from others. John Harris (F.R.S. 1696) read a paper on Leeuwenhoek's 'little animals in rain-water' and showed some examples at a meeting, while the still unknown Stephen Gray (only to be elected F.R.S. in 1733, in his old age) was introduced by Hooke to show a microscope of his own construction; his paper describing work he had done with it was read and published in the *Philosophical Transactions*. Hunt described his records of rainfall, foreshadowing the Society's careful maintenance of meteorological records in the eighteenth and early nineteenth centuries. Slare read a paper on chemistry, Hooke read one on petrifaction and one on the nautilus, which led to much discussion about creatures which might live in the depths of the sea, and there were medical papers, usually either clinical or anatomical. In the first half of the year Halley was, as before, very active, reading several mathematical papers: on mathematical optics, on mathematical astronomy, on conic sections; he also read two papers on classical geography and one on recoinage, the latter no doubt the result of his appointment to the Chester Mint, which deprived the Society of his presence in the autumn. No one replaced him as an active contributor until late in 1703.

Not that the Society totally lost its direction or its critical view of its function: for example, it is noteworthy that, when in January 1696/7 a paper on astrology was read, the meeting promptly voted to ask the author to produce 'some Proof of his Ability' (i.e. some successful predictions), after which nothing further was heard of him. There were no experiments performed at meetings and little in the way of reported experiment, except what was contained in letters. Leeuwenhoek provided good material for two meetings, Tyson read a paper on anatomy, and Lister read one on plant juices, so that the biological content was fairly high. Hunt, as in the previous year, summarised his rainfall data. Hooke read an account of a treatise on amber (by P. J. Hartman) which led him to write a considerable paper, occupying three meetings, about the origin, nature and characteristics of amber. Otherwise, Wallis produced a mathematical paper, Sloane read a number of more or less empirical letters, not particularly profound in nature, and there was much natural history and many accounts of books.

This relatively lively year preceded a fairly disastrous one. In 1698 attendance was generally poor, there are no minutes before 19 January nor between that date and 23 February (so presumably no meetings) and much of what meetings there were seem to have been taken up with 'curiosities', both plant and animal. On the positive side Gray once again showed a microscope designed by himself, Hooke described an anatomical dissection, and Woodward showed the results of examining a strange stone. Of genuine experiments shown there was only one practitioner, Geoffroy,[31] who showed several chemical experiments: on the Bononian stone (a 'phosphorus'), on odourless liquids which when mixed produced a fetid smell, and on the fumes produced when spirit of salt was poured on steel filings, which, as Boyle had done long before, he showed to be inflammable. A letter from Leeuwenhoek provoked good discussion, and there were agricultural accounts, one curiously by Fatio de Duillier. Many letters were read, but these contributed little or nothing to the experimental content of the meetings.

The reasons for the decline are not apparent, other than Halley's departure, for 1699 was considerably livelier in all respects, with more instruments shown, more (though still few) experiments shown or described, and more scientific papers read. Among the instruments may be noted a model and drawing of Savery's steam engine in improved form (25 June) and Newton's sextant, which he showed himself (16 August). John Lowthorp[32] read his 'observations concerning the rarefactions of the Atmosphere', which contained an account of experiments performed by him; he was then asked to make further experiments, which he did a month later before the Society with Hunt's assistance. Botanical experiments were described by Woodward; Lister read a paper on the physiology of the flea; Hooke read again on petrifactions; and Mr Bridgeman[33] presented a paper on minerals. There were letters from Leeuwenhoek and Geoffroy and accounts of experiments performed at the Académie Royale des Sciences. Thus at most meetings in 1699 there was some experimental content.

But, as often, a good year was followed by a poor one, and in 1700 experimental interest was generally low. Hooke read a paper on longitude determination, Papin sent in a highly experimental paper which was received with marked interest, Geoffroy sent a paper whose principal experiment, on the production of cold by a mixture of liquids, excited so much interest that it was, almost uniquely in this year, tried before the Society, probably by Hunt. There were a number of experiments described as performed elsewhere: Hunt spoke of trying to burn a piece of pork kept many years in a well (it behaved like fresh meat); Slare reported that he had confirmed that the weight of rocksalt increased when sea water was added to it, a fact stated by William Cockburn (F.R.S. 1696) but denied by a Mr Seaforth; and George Moult, apothecary (F.R.S. 1689), read a paper on evaporating sea water. Gray sent a paper on microscopes, and Leeuwenhoek's work, which provoked discussion, was read over four meetings.

In 1701 only a very few individual experiments were performed, but more than in the previous year and there were more empirical papers read. 'Mr Godfrey', almost certainly Boyle's assistant in the 1680s, showed the extreme inflammability of the 'fume' produced when oil of vitriol was poured on steel filings (as first perceived by Boyle); Abraham de la Pryme (best known as an antiquary) described his experiment of steeping grain in water to improve germination; and Halley spoke on weather and showed his chart of the English Channel. A surgeon, William Cowper (F.R.S. 1699) showed some anatomical preparations and read a medical paper. Sloane read papers sent in to the Society mainly on 'curiosities' but one on colouring marble and another with some experiments on bezoar stones. Hunt read a paper in which he compared the weights of lead and lead ore, Gray sent in a letter describing a new hour glass, and there was an Italian paper on Indian varnish. There was an important and experimental paper on 'A Scale of Degrees of Heat' which so pleased the meeting that it was ordered that it be printed forthwith, still anonymously. (Did some present know or guess from its experimental tone that it was by Newton? It is indeed worth printing and contains Newton's Law of Cooling implicitly although not explicitly.) Other subjects of continuing interest were natural history and medical case histories, concerns certainly empirical although hardly experimental. Hooke read only once in 1703, a review of Huygens' *Cosmothereos* (published 1698); this was to be Hooke's last recorded contribution before his death in March 1703.

In 1702 and most of 1703 interest and activity, which was not very high, centred on physiology, anatomy and microscopy. Cowper again showed anatomical preparations, and performed an experiment in physiology, and also read several reports of his work. The letters of Leeuwenhoek were read slowly, occupying many meetings. Richard Richardson (F.R.S. 1712), physician and botanist, sent in drawings of fossils, Sir Charles Holt sent letters on microscopy, and John Martyn provided an account of a clock of his own design. Halley was often present in 1702, and read a paper on cometary motions. Hooke occasionally attended but, as noted above, was mostly inactive; when he died, the only immediate change which occurred was the strengthening of the threat begun in 1702 of eviction of the Society from its rooms in Gresham College (a threat which hung over the Society but was not, by force of circumstance, to be executed for a good number of years).[34]

But changes were to occur. The election of Newton as President at the Anniversary Meeting of 1703 was followed soon by the appearance of Francis Hauksbee, who on 15 December produced his airpump and proceeded to begin a series of experiments which altered the character of the Society's meetings for the next ten years.

~ 6 ~

The communication of
experiment
1677 ~ 1703

With Oldenburg's death, the Society was faced with a crisis in communication with which it was ill equipped to deal. In the autumn of 1677,
although some letters continued to come in from abroad (naturally still
addressed to Oldenburg), no one, not even Henshaw who was officially
Secretary, seemed prepared to answer them, while the *Philosophical
Transactions* (after all, Oldenburg's private business) came to an abrupt
end. Nor was there immediately anyone prepared to take up such tasks
as those of preparing Malpighi's works for publication. No future
Secretary was ever to be quite so zealous as Oldenburg until Sloane
became Secretary in 1693; he was very energetic, but busy though he
was he did not have quite so many tasks in hand as Oldenburg, nor was
he so skilled as Oldenburg in establishing a network of correspondence.
Although the Society was to take such steps as it could to re-establish the
tried methods of communication, it took time. Not everyone was at first
convinced that foreigners and strangers ought to learn of the Society's
affairs, and future Secretaries had to be both guided and goaded in their
tasks.

The first concern of the Council was to recover the Society's books
and papers which had been in Oldenburg's hands, a task complicated
by the fact that Oldenburg had died intestate, as well as by Hooke's
eager desire to peruse all papers relating to the controversy between
Huygens and himself, which, he was certain, Oldenburg had managed
to his, Hooke's, disadvantage. The second task was to regulate the
Society's affairs in a period when the President, Brouncker, was ill, and
the surviving Secretary, Henshaw, was acting as Vice-President in his
place.[1] The Council made a number of orders when it met on 24
September, two weeks after Oldenburg's death, of a general kind: that
Secretaries should communicate the contents of letters to the Society as
soon as possible after their receipt, and that the 'Officiating Secretary'
should read out at each meeting his 'short notes of all that passes at the

Society or Council'. But not until nearly a month later (18 October) was Hooke appointed as Secretary in Oldenburg's place; he was then formally elected at the Anniversary Meeting, with Grew as the other Secretary and Sir Joseph Williamson as President.

Hooke began his term of office by reading a brief note from Leeuwenhoek, dated 6 October, which he had addressed to Brouncker as President, having by then learned of Oldenburg's death.[2] The accompanying paper Hooke read in full to the meeting a fortnight later, whereupon he 'was ordered to return the Society's thanks to Mr Leeuwenhoek, and to endeavour to procure further discoveries from him by holding correspondence with him'. This he did on 30 November,[3] eliciting a prompt reply; it must have been confusing to Leeuwenhoek that this time it was Grew who responded, Grew having now taken over much of the Society's correspondence as well as the taking of the minutes. Another who assisted Hooke in correspondence was his admiring friend John Aubrey, who had much to do with Newton's correspondence on optical matters with Anthony Lucas.[4] Someone, no doubt Grew, also proceeded to put in hand a new Letter Book into which, as in the past, the more important letters received were, with their replies, copied in by the amanuensis; this was a task which Oldenburg had, unaccountably, neglected in the last months before his death.

Domestic correspondence was not a problem; all the Secretaries appointed after 1677 were reasonably receptive and cordial when provincial Fellows and virtuosi sought exchanges of letters, as both the archives and various printed letters amply show. But foreign correspondence was more complex, especially since the frequent change of Secretaries confused those who wrote infrequently, so that many foreigners were uncertain to whom they should send their letters. Also many were slow to learn of Oldenburg's death: even so regular a correspondent as Hevelius did not know of it until well into 1678, while the very last letter addressed to Oldenburg, admittedly from Indonesia, was dated 1681.

At the end of 1677 Grew drew up a formal Latin letter designed to be sent to Oldenburg's most important correspondents. Some of these – Malpighi, Huygens, Justel, Leibniz – were already aware of Oldenburg's death,[5] while it is not at all clear why the names of Petty, Lister and Newton were included, unless the letter was mainly designed as an official notice from the Society that regular correspondence was welcome. This letter was slow to be sent out, partly because Grew had to seek approval of its contents from the Council; in fact he dated his letter 31 March 1678, sending ten copies to Williamson for transmission abroad and promising another ten shortly, implying that the list of recipients had been extended.[6] As the Society discovered, it was one thing to urge savants to write and another to make them feel as much a part of the Society's affairs as Oldenburg had been able to do. As late as February 1679/80 Flamsteed wrote sadly[7]

> Our meetings at the Royal Society want Mr Oldenburgs correspondence and on that account we are not so well furnished nor frequented as formerly, but I hope a little time will put us into order.

This time was to be slow to do. It is true that domestic correspondence soon picked up, both from individuals and from other societies. And from 1683 to 1686, as the minutes show, the Oxford Philosophical Society provided much of the experimental entertainment at meetings, by means of detailed accounts of their activities sent by Musgrave and by Plot, both subsequently Secretaries of the Royal Society, while Aston, in turn, sent news of the Royal Society's activities. In filling their meetings with the repetition of experiment, the Oxford Society was obviously returning to the principles which had animated the Royal Society in its early days; similarly, the Oxford Society like the Royal Society flagged after a few active years, so that after 1687 little is heard of its work at Royal Society meetings.[8] The Dublin Philosophical Society, founded in 1683, also intended to fill its meetings with experiments; like others, its members maintained activity for a little period, in this case from 1683 to 1685, sending careful reports to London, but, although after a lapse of five years activity revived, it did not last long.[9] The Royal Society listened attentively to the letters sent by its Secretary, William Molyneux, mainly through Aston, with their detailed accounts of experiments performed in Dublin which provided good experimental content to the London meetings. But during these and later years much of the correspondence from provincials gave details of natural history or the antiquities found in their neighbourhoods, and these, by their nature, could contribute little or nothing to the experimental content of Royal Society meetings, although they usually were heard with considerable interest.

Foreign correspondence was slow to increase in the 1680s. Partly this was because no Secretary between Oldenburg and Sloane was internationally orientated, partly because the linguistic equipment of most Secretaries was poor (Waller was apparently an exception) so that Latin correspondence, such as that with Hevelius, Cassini and Malpighi, was easiest to handle, partly because the Secretaries apparently had no ability in initiating correspondence such as Oldenburg had possessed in abundance. During 1678 most foreign correspondents wrote to Grew, who had sent out the formal letter inviting correspondence, as Huygens did in framing a polite letter of congratulation to the newly elected officers of the Society after the Anniversary Meeting of 1677, although some addressed the President and others 'the distinguished, learned, eminent men of the English Royal Society'.[10] When Grew was replaced by Gale in November 1679, many correspondents addressed their letters to Hooke. Gale, a classical scholar, not a natural philosopher, was relatively unknown abroad, and, although he took over some correspondence in 1680 and 1681, it was mainly under others' direction. When Aston, equally unknown abroad but at least a virtuoso, replaced Hooke in November 1681, he relieved Hooke and successive Secretaries of much of the burden of foreign correspondence until 1685. Some found their own routes: that inveterate French correspondent Justel, who had written weekly to Oldenburg over many years, addressed his letters in 1678 and 1679 to Hooke, subsequently shifting to Halley as recipient, especially after he moved from Paris to England in 1681 to avoid persecution in France as a Huguenot.

The problem of how to address the Royal Society during these years is clearly seen in Leeuwenhoek's methods, evolved after some years of confusion. Having had the letter which he had sent to Brouncker in 1677 answered by Hooke (who had it seems relied on Haak for the translation of Leeuwenhoek's Dutch), he continued to address Hooke until he ceased to be Secretary. (To confuse Leeuwenhoek further, it was Gale who signed the formal notice of his election as F.R.S. in 1680.) He then tried addressing Wren as President, only to receive a reply from Aston. So he continued to address Aston until the latter ceased to be Secretary in 1685. In that year Halley became Clerk and began to write letters at the Society's orders; this further confused Leeuwenhoek, who thought the letter must have really been written by Gale, with whom he had exchanged letters five years earlier.[11] After this, seeking the only practical way out of confusion, he simply addressed his letters to 'The Royal Society', taking the precaution of having the most important translated into Latin in Holland. He maintained his practice of sending his discoveries to London in spite of the fact that, apparently, he heard nothing at all from the Society between 1686 and 1691, possibly because of Halley's lack of interest in biological matters. In the 1690s things improved again: in 1691 William Stanley (F.R.S. 1689), who had met Leeuwenhoek when he was chaplain to Queen Mary before 1688, wrote to present the Society's greetings and respects, while in 1692 Waller, as Secretary, resumed correspondence by informing Leeuwenhoek that his latest letter had been read at a meeting of the Society and, as always, had provoked much interest. He then continued the correspondence through 1694. But, although Waller and Sloane, not joint Secretaries, must have realised that Leeuwenhoek's letters had, over the past fifteen years, provided much important experimental entertainment at Society meetings, neither thereafter proceeded very zealously. As Leeuwenhoek's letter of 19 February 1697 (N.S.) makes clear, Sloane had written (8 December 1696), only at the Society's order, to convey news of the Society's recent publishing activities. Leeuwenhoek responded eagerly, requesting copies of the *Philosophical Transactions* and of Malpighi's recently published *Opera Posthuma* and promising to send his printed Latin letters. This he soon did (10 September 1697), acknowledging the receipt of the publications he had requested. Later letters specifically ask for the Society's reactions to what he had sent, suggesting powerfully that Sloane had been less than punctilious in keeping him informed, although the letters had certainly been read at meetings and commented upon. In spite of this, he continued to send the Society accounts of his discoveries until his death in 1723.

Leeuwenhoek's experience is almost certainly typical of that of many foreigners, for, whatever he may have thought, his lack of English and Latin was not a real barrier to understanding, when there were so many Dutch speakers in England, and especially in London.[12] And indeed almost the same experience was partly shared by Malpighi, who had no language barrier to overcome since he always wrote in Latin. Like Leeuwenhoek's his works were always favourably received and deemed worthy of publication in England. It seems rather to have been that frequent changes in Secretaries and confusion between the two current office holders prevented any stable system developing. Apparently, Hooke was a fairly active foreign secretary, especially in his first years, although he obviously preferred other rôles. Aston was generally more industrious in every respect for his four years in office. Then in 1685, shortly after the Anniversary Meeting at which he had been chosen again, he suddenly resigned. The minutes are discretely silent about the reason, but it is probable that he complained of overwork, lack of assistance (there had been no Clerk during the previous year), and too little remuneration. Hence, presumably, the decision, slowly implemented, to appoint a Clerk more highly skilled than Wicks, who when dismissed in 1683 had served for twenty years. Now Halley was chosen, after which the Secretary's job became less arduous.[13] Until then, correspondence was disrupted, to thrive only after Sloane became Secretary in 1693.

But even Sloane, with all his acquaintance with the learned world abroad, could not form such a network of correspondence as Oldenburg had done. Such men as Leeuwenhoek, Malpighi, Hevelius and Cassini continued to write as they long had, but their letters were less frequently read at meetings than had been the case in earlier decades, nor did they receive as much English news in reply as they had done in Oldenburg's time. Even Sloane does not seem to have been able to incorporate into his letters a clear picture of what was occurring at home and abroad, so that there was less sense of a community of learning in which all correspondents shared. Nevertheless, in the 1680s and 1690s the work of the Secretaries and the Clerk did serve, even if somewhat haphazardly and intermittently, to generate that communication of experiment to which the Society was dedicated.

The problem was recognised by successive Secretaries, and the failure to procure much philosophical news from at home and abroad was blamed for a decline in the Society's affairs. Thus in 1687 Tancred Robinson, Secretary for a fortnight before Aston's sudden resignation in 1685, told Sloane, then in Jamaica,[14]

> The Royal Society declines space, not one Correspondent in being. The Revenue is settled upon Mr. Hook, and Monsieur Papin goes next week to settle in Germany. The same officers are chosen this year; I am afraid that you will find nothing but ruins on your return.

This was the situation which Waller tried to resolve by picking up the threads of, particularly, biological correspondence, in which Halley (in any case very busy during the previous year in seeing Newton's *Principia* through the press) was not interested. Waller was also active in trying to increase correspondence on such subjects as natural history, as when he wrote, probably in 1693, to a great traveller, the Reverend John Clayton,[15]

Rever'd Sir

finding several ingenious Papers of the curious and usefull Discovery's & Observation made in Virginia I have taken the liberty granted me by the Royal Society of inserting what I thought fit of their Papers in the Transactions, which I have for some time undertaken to publish; accordingly have printed one or two of your letters therein. My Request therefore to you is that you would please to continue your Communications, and go through with your Natural Acct. of the Countrey as to the plants and Animals which in one place of your Letters you intention, which will oblige your friends of the Royal Society. and If I can serve you in any thing here, you shall not fail in a Correspondent in Sr Yours &C

As this letter suggests, Waller's interest lay in correspondence which would furnish material for the *Philosophical Transactions* (see below) while other correspondence he generally left to Halley or to Sloane. Halley, who had maintained an active but selective correspondence, concentrated on astronomical matters, on the whole writing only as directed. In 1694 he was absent for the whole of the summer, leaving his duties in the hands of Sloane (who had, ironically, been a rival candidate for the Clerkship in 1685), and soon he left London altogether. When that happened, he nominated Dr John Arbuthnot (F.R.S. 1704) as his deputy. Arbuthnot, physician and mathematician, served until the end of 1698, when he was succeeded by Jezreel Jones for two years, followed by Humphrey Wanley (F.R.S. 1706).[16]

Sloane, well provided with clerical assistance, was eager to take up both the Royal Society's domestic and foreign correspondence, soon taking over nearly all of Waller's correspondents. Like Waller, he began to write to prospective correspondents solliciting an account of their work, in a form showing his private interests and without any reference to experiment. (But probably this was a letter intended to be sent widely and to serve many purposes.)[17]

The Royal Society was resolved to prosecute vigourously the whole design of their institution, and accordingly they desire you will be pleased to give them an account of what you meet with, or hear of, that is curious in nature, or in any way tending to the advancement of natural knowledge, or useful arts. They in return will always be glad to serve you in anything in their power ...

The most important and novel increase in correspondence in the later 1690s was that with France, which had been in serious decline a few years earlier, when Oldenburg's correspondents were growing old and the Académie Royale des Sciences, in the years preceding its thorough reform of 1699, was decidedly moribund. In 1698 Lister, spending some time in Paris, found the Académie seemed to have very few members and no power to appoint more. Those left told him they longed for news of what was happening in England for, as Lister remarked, 'the Wars had made them altogether strangers to what had been doing in England'.[18] The Marquis de l'Hôpital, a member of the mathematical section of the Académie, proved to be very favourably disposed towards England, longing for back copies of the *Philosophical Transactions*, and pleased to be informed of what Lister called Newton's Preferment (appointment as Warden of the Mint in 1696) and 'that there were hopes, that they might expect something more from him'.

The chief contact with France came from Sloane, who had become intimate with the physicians of Montpellier and with the Parisian botanist Tournefort and had kept up an intermittent private correspondence with them. Now, active in the Society as Secretary, he found much in their letters suitable for reading to the meetings. Admittedly, by no means all of these letters contained experimental information, especially since, as Sloane's letter inviting correspondence suggests, his own interests centred on natural history, his extensive botanical knowledge being directed either to the understanding of new plants and their classification or their use in medicine. Sloane's greatest contributions to the provision of experiment for the Society's meetings came less from physicians and botanists than from his meeting with the chemist E. F. Geoffroy, an active member of the Académie Royale des Sciences before and after its reorganisation.[19] Geoffroy came to England in 1698 with a letter of introduction to Sloane provided by Tournefort. Sloane quickly proposed Geoffroy for election to the Royal Society, after which he attended several meetings, at some of which he performed experiments (see Chapter 5). In years to come, Geoffroy continued his correspondence with Sloane and frequently described experiments performed in Paris, thereby giving a valuable experimental orientation to Sloane's French news which naturally otherwise inclined to the botanical or medical. And when Sloane was elected as a 'correspondant' of Jacques Cassini by the Académie des Sciences, he was further put in touch with the world of French physical science.

As in Oldenburg's time, there was in the 1680s a considerable amount of Royal Society sponsorship of domestic and foreign experimental treatises. Oldenburg had since 1669 seen Malpighi's books through the press, editing them with care. Now Hooke did the same for Malpighi's

Anatomes plantarum pars alterum; like the earlier works it was dedicated to the Royal Society, to whose meetings most of it had been read in summary. In 1684 Malpighi's 'Dissertatio epistolica varii argumenti', addressed formally to Jacob Spon, was printed by Plot in the *Philosophical Transactions* (Volume 14), while his 1689 work, *De Structura glandularum conglobatorum consimiliumque partium epistola* was sent by the author to Waller, who presumably saw it through the press. Malpighi's *Opera omnium* of 1686 was a private enterprise by several London booksellers and printers, which was not sponsored by the Royal Society, although Malpighi may have thought that it was; in any case he was pleased by its appearance. But when the *Opera Posthuma* was sent in 1696 by Silvestro Bonfiglioli, friend and associate of the now deceased Malpighi, Bonfiglioli was promptly elected F.R.S. and the Council charged Hoskins, Hooke and Sloane to arrange for its printing, as they did.[20]

Malpighi's work had always appealed to the Society, typifying as it did the glories of the experimental approach which the Royal Society never failed to promote. Hoskins, President 1682–3, and often Vice-President, was particularly energetic in promoting publication and translation of experimental work by foreign natural philosophers. It was Hoskins who finally arranged for the publication of Malpighi's *Opera Posthuma*, and it was Hoskins who had encouraged Waller in 1684 to translate the Accademia del Cimento's *Saggi* of 1667 into English under the title *Essayes of Natural Experiments Made in the Academie del Cimento*, and saw to its licensing by the Society. Waller continued to be active as editor and translator during his Secretaryship: in 1702 he published a translation of Picard's *Mesure de la Terre* of 1672, no doubt assisted by the summary which Oldenburg had published in his *Philosophical Transactions* of 1675, while in 1705 he was to be responsible for publishing Hooke's *Posthumous Works*.[21] He was in the 1680s one of those appointed to the Royal Society's committee charged with the publication of Francis Willughby's *Historia piscium*, to be effected in 1686. This work, compiled by Ray after Willughby's death in 1672, partly from Willughby's manuscripts, partly from his own work (it was always referred to in the Society's minutes as being by both men), was a very expensive work to publish because of its lavish illustrations. Its publication would never have been undertaken without the activity of Lister, then a Vice-President, who in 1681/2 urged it on the Society, nor without Robinson's urging of Ray to complete it.[22] The result was a handsome book but, as is well known, a financial disaster for the Society, so much so that

Halley's salary had to be partly paid 'in fishes', and it was proposed that Hooke's should be also (he declined). This is the reason for the Society's failure to give financial support for the publication of Newton's *Principia* to which only moral support and a license (by Pepys as President, dated 5 July 1686) could be given. No wonder that the Society declined to sponsor any other work, nor that such sponsorship was left to individual Fellows.[23]

The Society did not support the *Philosophical Transactions* (any more than it ever had), although it urgently longed for its resumption. The last issue edited by Oldenburg had been no. 136 (part of volume 12) dated 25 June 1677; he had blamed the printers' holidays for his failure to publish any issues in the summer of that year. In any case, his death naturally put an end for the time being to what had been, after all, his private journal. But it was sorely missed: four months later (2 January 1677/8) the Council in debate favoured continuing the journal, except for Hooke who argued against the proposal, although on 7 September 1677 he had been carefully completing his set and on 8 February 1677/8 was pleased to obtain the index. According to Hooke, Grew was eager to continue editing the journal and was supported by Henshaw. What was decided, according to the Council mintues of 2 January, was

> That there be prepared once a-year a collection of all such matters, as have been handled that year, concerning, four, five, or more subjects, which have been well prosecuted, and completed; which may be printed in the name of the Society against the anniversary election day.

And, also,

> That the Register-books of the Society be perused; and that what shall be thought fit by the Council to be published, be drawn out and published accordingly.

The first resolution was never implemented or mentioned again. As for the second, it must be related to Grew's desire to continue the *Philosophical Transactions*, much to Hooke's dissatisfaction. Hooke in his *Diary* for 24 December complained that Grew had been at Boyle's 'solliciting for Transactions', while two days later he remarked 'Henshaw a good man but for Transactions'. He tried to block the journal's revival, but, although at the end of the meeting of 2 January he could note 'I moved against Transactions, brought them to be of my opinion', in fact he was finally defeated, and Grew published his first *Philosophical Transactions* (no. 139) as of 10 February 1677/8. This consists almost entirely of material drawn from the archives, in conformity with the Council's resolution, together with book reviews, this archival material consisting mainly of histories of trades. Grew completed volume 12 of the *Transactions* a year later with numbers 140–2.

Although Hooke had worked so fiercely to prevent the resumption of the *Philosophical Transactions*, he was not completely opposed to the publication of new material, even that by foreigners: by the time that Grew published his first number of the *Transactions* there was available *Lectures and Collections Made by Robert Hooke, Secretary of the R. Society*; this, dated 1678, contained two works, *Cometa* and *Microscopium*, the first including papers and letters by Boyle, Halley and Cassini together with extracts from the *Journal des Sçavans* as well as several papers by Hooke himself on comets, the second letters by Leeuwenhoek and some commentary on microscopy. Similarly, Hooke's famous Cutlerian Lecture *De Potentia Restituva, or Of Spring*, also published in 1678, has as a supplement 'some Collections', that is, papers by Papin, by James Yonge, a naval surgeon, several papers on natural history with comments by Hooke, and finally an account of a volcanic eruption in the Canary Islands. Both these works are almost as miscellaneous as the contents of the Society's meetings and like them pleased Hooke by giving him the opportunity to make comments on the work of others. It was perhaps this which displeased him about the publication of such collections by others, that is, that it did not permit him to arbitrate and comment. And he obviously disliked, vehemently, the very name of Oldenburg's journal.

For when volume 12 was completed by Grew, the Council would have liked either Grew or Hooke to continue it. Hooke fought a delaying action but finally (by 3 July 1679) agreed to publish what the Council continued to call the *Transactions* but which he was to re-name the *Philosophical Collections*. The first issue was dated 1 November 1679, although it was given to the printer early in October.[24] This really differed from the *Philosophical Transactions* in name only, for every number (ultimately seven in all) contains short articles, English and foreign either excerpted from letters or taken from printed sources, together with short notices of books. The Council wanted this to continue: on 8 December 1679 in a review of policy it resolved

> That the Secretaries take care to have a small account of philosophical matters, such as were the Transactions of Mr Oldenburg, and under the same title, published once a quarter at least, and that it be recommended

to them to do it monthly if it may well be; but at least that it be done quarterly.

When Hooke was asked if he would agree he replied 'that he would see what he could do in it, but could not as yet undertake it absolutely'. And, although he discussed and wrote about the possibility in the spring of 1680, he did nothing for another year.[25] At last in 1681 he complied with the spirit but not the letter of the Council's resolution by producing two numbers of his *Philosophical Collections*, following this with four more numbers in the first four months of 1682. He then ceased, and no more numbers appeared; he settled the question further by resigning the Secretaryship at the end of 1682.

It seems to have been Aston (who had become Secretary in November 1681) who realised that with Hooke no longer a Secretary, it was possible to resume publication of a journal under the old, familiar title of *Philosophical Transactions*. He opened negotiations with the printers promptly, as appears from the minutes of the Council meeting on 13 December 1682 when it is recorded that

> Mr Aston, having acquainted the Council, that the Philosophical Trans-
> actions might be carried out the next year, if some encouragement was
> given to the publishers, by taking off a good number of copies, as soon as
> they sould be printed

the Council agreed that the Treasurer should buy sixty copies of every issue 'for the use of the Society'. (The publishers were, of course, not the printers, but the editors.) Oldenburg had hoped to make money from his journal, but had never done so; now, after so irregular a publication record, it was clearly reckoned that any Secretary undertaking publication needed a guaranteed sale. The official editor was Plot, working from Oxford; Aston was known to be helping him, and Tyson gave some assistance; there was also at first the unsatisfactory 'G.' of Aston's letters to Plot who insisted upon anonymity; he was probably not a Fellow, but a clerk or amanuensis who dealt with the printers. The next two volumes were printed at Oxford. Very sensibly, no. 143 of January 1682/3 began with a disclaimer insisting that the editor, not the Society, was responsible for publication.[26] This was coupled with an appeal for contributions and directions for submission of them. 'Although the Writing of these Transactions', the preface begins, 'is not to be looked upon as the Business of the Royal Society', yet 'they are a Specimen of many things which lie before them', they are important for the world at large to know and they preserve 'many Experiments' which would otherwise be forgotten. Plot went on to hope that those who have appreciated them

> will most readily endeavour, themselves or by others, to supply and keep
> up that Stock of Experiments, and other Philosophical Matters, which
> will be necessary hereunto.

The *Philosophical Transactions* did continue to appear more or less monthly for the next four years, edited first by Plot and then, for one year, by his successor as Secretary, Musgrave, after which the newly appointed Clerk, Halley, took over. In Halley's issues there is a paucity of biological papers (none by Leeuwenhoek) and a preponderance of astronomical papers, reflecting his own interests. Curiously, Halley discontinued the effort of publishing the journal after 1686, at the very time when his editorial duties for Newton were about to cease. Once again, the *Philosophical Transactions* disappeared.

It was not until late in 1690 that the Society seriously endeavoured to revive them. This was not, as usually stated, under the influence of Southwell, elected President on 1 December, but rather of Hoskins, who as Vice-President presided at the Council meetings at which the necessary decisions were taken. First an attempt was made to interest Hooke again by agreeing to reimburse him for postage on the letters he had written and received on the Society's behalf as Secretary 'on condition that he publish Transactions or Collections as formerly'; the Society was willing 'to take' (pay for) sixty copies as before. But Hooke evidently refused and the matter was deferred for a full Council meeting at which Halley could be present, as happened on 28 January 1690/1. Then

> it was resolved that there shall be Transactions printed and that the
> Society will consider of the Meanes for effectually doing it.

To this end, Tyson, Slare, Sloane, Waller and Hooke were all asked to assist Halley 'in compiling and Drawing up the Transactions', while Boyle was asked to continue as before to send 'his small Tracts' for publication. (But he was to die at the end of 1691.) Clearly the intention was that Halley should continue to edit the journal but now with help in order to make his burden less. But in the end it was Waller who agreed to 'compile and draw up' the next two volumes (17 and 18), although slowly (it took him over three years). Waller gave the explanation for the hiatus in publication as being

> chiefly by reason that the unsettled posture of Publick Affairs did divert
> the thoughts of the Curious towards matters or more immediate concern
> than are *Physical* and *Mathematical* Enquiries,

these latter being, as he said, the peculiar concern of the Royal Society.[27]
He concluded his prefatory 'Advertisement' by solliciting contributions
of 'Discoveries in Art or Nature' which, he promised, 'shall be inserted
herein, according as the Authors shall direct', an attitude different from
that of his predecessors, who emphasised their own judgement, choosing
from what was available with some discrimination, rather than letting
potential authors dictate the manner of publication, but continuing
what had by now evidently become the custom. Would-be authors were
to send their contributions to Henry Hunt rather than to the publisher,
again suggesting that they could be certain of publication. In spite of
this, Waller did not soon achieve regularity of publication; he managed
only three numbers in 1691, while number 195, dated 1692, seems not to
have been printed until March 1692/3. This may explain why the
Council began to consider other possible editors. On 7 December 1692
the Council recorded that

> Dr Halley offered that it should be undertaken to print a book of
> Philosophical matters such as the Transactions used to consist of, that he
> would undertake to furnish de proprio five sheets in Twenty.

Hooke, whose friend Waller was, evidently saw this as a move against
Waller and a criticism of his work, writing in his *Diary*,

> A councell of Royal Society Hosk[ins] and Hill; shuffled of[f] Plot and Mr
> Waller, and made Hally to be Secretary. None els sayd anything, soe
> councell brake up, nothing done: noe *Transactions*, noe correspondence.

(It was of course not true that Halley was made Secretary, nor that
Waller and Plot were at this time displaced.) Then on 15 February
1692/3, 'It was resolved that Dr Plott shall print the Transactions' with
the same financial inducements that had been offered to Halley.[28] But
this also was not implemented, and on 17 February Hooke recorded 'At
Wallers: he will correspond and publish transact', a decision he
confirmed to Hooke on 28 February.

Waller did indeed now resume both his Secretarial duties and his
publication of the *Philosophical Transactions*, completing volume 17 and
publishing volume 18. Waller wrote a significant preface to no. 196
(dated January 1692/3)[29], which both emphasises the private nature of

the journal and his intention to promote the experimental side of the Society's design. Like Stephen Hales later, Waller declared that the application of '*Number, Weight* and *Measure*' to the 'Problems of Nature' could alone lead to an understanding of the natural world, adding that

> This has been the *Employment* of the *Experimenting* part of Mankind, and the *Design* of that *Glorious Institution* of the *Royal Society*.

This sentiment of course not only anticipated the spirit of the eighteenth century but mirrored that of nearly half a century earlier and the permanent aim of the Royal Society in its first three-quarters of a century. And Waller worked hard to live up to this aim and to secure material for his *Transactions* which should accord with it. He encouraged both domestic and foreign correspondents to send him copy; at least once he had an anonymous contribution refereed; he seems to have tried to maintain a balance within each number of the different subjects available to him; and he undertook to provide many of the illustrations.[30] He was obviously personally much interested in Leeuwenhoek's contributions, publishing them regularly and raising points of interest in his letters to Leeuwenhoek, being especially taken with those relating to generation. In accord with the Royal Society's early aims, he was eager to publish histories of trades. He was obviously a conscientious and industrious editor. But he was not able to publish monthly numbers, partly because of difficulties with the printers. Hooke recorded a copy of no. 196, dated January, as received only on 29 March, that of no. 197 as received on 12 April, while 'Mays transactions' he did not have until July,[31] although June's number was available on 28 July. To a certain extent Waller overcame the worst of his difficulties, finally settling for bi-monthy numbers, a practice continued for a couple of years by Sloane after 1695.

Whether Waller was pleased to give up the publishing of the *Transactions* is not known. Sloane, in his first number (January/February 1694/5), was so ungenerous as to imply that there had been a hiatus of several months in the production of the journal, which was not really true, while at the same time admitting that 'the Royal Society commanded me, sometime since, to take care to continue them'. Sloane was to remain the publisher for the next twenty years, that is, during most of Newton's Presidency. He intended to follow Waller's practice of endeavouring to produce a well-balanced journal; that, at least according to the satirists, he often failed to do so must be in large part because of the inadequate scope of his correspondence. In the Preface to his volume 21 for 1699 he declared that it was his custom to include papers of varied types: some might contain 'Matters of Fact, Experiment or Observation', while others were concerned with 'Hypothesis'; as to these latter, he remarked drily, they 'may be pass'd over by such as dislike them', as he implied that he did himself. He was to be a careful editor as regards details, for example always identifying writer and recipient of letters although not yet, as was to be the custom in the next century, indicating when they were read to the Society. Inevitably, given his interests and his range of correspondents, he leaned heavily towards the biological side and, as in the letters he read at meetings, so in those he published, there is a good deal of now seemingly trivial natural history, including curiosities. There are also many papers of antiquarian interest, these in keeping with the trends of the 1690s when the composition of the Fellowship changed somewhat and there was more interest by some older Fellows like Hooke in this sort of subject, which continued to occupy the Society even after the founding of the Society of Antiquaries some decades later. But Sloane never neglected physical science when it came his way and published many letters on 'hard-line' physical science during his editorship. Indeed his *Philosophical Transactions* remained a valued reflection of the Society's interests, including its experimental interest, and was so regarded by readers at home and abroad.

The *Philosophical Transactions* in large part mirrored the day-to-day interests of the Society, although, given that it was still the private property of its editors, even though they were also Officers of the Society, it was necessarily also a reflection of individual tastes. And these men gradually represented not the Royal Society of the 1660s but rather the Royal Society of the 1680s and 1690s. In 1680 the Society was barely twenty years old; nevertheless, many of the men who had founded it and had helped to shape its empirical tradition were now dead or at best aging and relatively inactive. Wilkins had died in 1672, Goddard in 1675, while Petty, Croone, Brouncker and Seth Ward were all to die in the 1680s, their best work done earlier. Wren was no longer concerned with mathematics and natural philosophy. The tone of meetings reflected this.

Among the founders, Boyle (1627–91) was still reasonably active. He had never lost his dedication to experiment and still sometimes sent in accounts of work done in his laboratory. But, mainly through ill health, he had long ceased to attend meetings with any regularity. However, he continued to publish new work throughout the 1680s, partly with the aid of assistants, like Papin and Slare, who in turn absorbed his devotion to experiment and served the Society well in the presentation of varied experiments. Boyle was to continue to be highly respected, above all for his experimental natural philosophy at home and abroad, and his works continued in demand into the eighteenth century. Papin had as Curator made signal contributions to the work of the Society in the 1680s, and had developed a Continental reputation which helped to publicise the Society's aims abroad. Slare, an important contributor to the experimental content of the Society's meetings and to the *Philosophical Transactions*, contributed to the Society's domestic reputation.

Hooke, first as Curator, then as Secretary, before becoming, sometimes, Curator again, played a dominant rôle after 1677 in directing the structure of meetings, but he published less and less after 1678, abandoned serious astronomy and indeed, in the 1690s, most experimental science as well. To other Fellows his work was a model of one sort of experimental learning, his *Micrographia* widely acclaimed by both serious natural philosophers and by virtuosi, but it was never reprinted nor translated into any foreign language in his lifetime. Hence Hooke's reputation abroad could not compete with that of Boyle or Wallis, and he wrote very little after 1680. His greatest influence in fact was in some ways after that time and on virtuosi like Waller, to whom he seemed the epitome of what the Royal Society had long stood for, and who could not in any case understand his more serious theoretical contributions. (Of course, his greatest influence had been on Newton in the late 1660s and in the 1670s.) By the 1690s Hooke, like many aging scientists, had become interested in subjects far removed from his greatest contributions.

Of the original founders, only the long-lived Wallis (1616–1703) continued to be active to, and even a little beyond, the end of the century, continuing to send letters to the Society, some even on empirical subject, many of them to be published in the *Philosophical Transactions*, while the publication of his *Opera Mathematica*, in three large volumes (Oxford, 1693–9) brought to the public at home and abroad not only a life-time's work in mathematics and mathematical physics but recalled the brilliant days of the late 1650s with the work of his disciples, Wren and Neile. Among astronomers after 1675, the most important were Flamsteed and Halley, the first, appointed Astronomer

Royal in 1675, excelled as an observational astronomer, the second rather as a theoretical one, although Halley did publish papers on experimental physics, particularly magnetism, as well. The botanical works of John Ray and the empirically based writing on entomology and fossils of Martin Lister were important for the Society's domestic reputation and were certainly models of what the Society desired, but they did not have a great influence outside England.

Clearly the only natural philosopher with a reputation in the learned world which could match that of Boyle, and the only one to uphold as vigorously as Boyle the ideals of experiment as a firm basis for theoretical natural philosophy, was Newton. His optical papers had not only demonstrated his profound belief in the major rôle which experiment ought, he believed, to play in the investigation of nature, they had implied that this too was the belief of the Royal Society. The Society's sponsorship of his *Principia* associated his name and his work with the Society once more, even though he had virtually withdrawn from it in the years following Oldenburg's death. And although his strong dislike of Hooke kept him from contributing more than very seldom in the 1690s to the Society's activities, he had not ceased to do so completely, and the Society continued to try to draw him back by appointing him to its Council. Outside the Society's circle, he yet remained an important member of it, and so, obviously he was seen by the Society itself, anxious to link him ever more closely to itself, and only awaiting a convenient opportunity to do so.

~ 7 ~

The record of the minutes
1703 ~ 1727

With Hooke's death there was to be a very considerable change in the Royal Society's activities. This was not quite the same situation as had arisen in 1677 when the death of a very active Secretary demanded as well as permitted change; it was rather that the death of Hooke, a former Secretary and more or less perpetual Curator of Experiments, *de facto* if not *de jure*, since 1662, made possible the installation of a new President (for Newton would never have accepted election while Hooke was alive),[1] and this in turn led to the appointment of a new Curator of Experiments. Newton was to play an active part in the Society's affairs, presiding over the vast majority of meetings, commenting on, occasionally presenting experiments, and above all helping to guide the presentation of experiments, now far more numerous than had been the case in preceding decades, for the next fifteen years. After this, in Newton's extreme old age, the momentum generated previously then continued for another decade. The presence of a new Curator of Experiments (not always so called) totally revolutionised the Society's meetings. Francis Hauksbee, who suddenly appeared at the Society in December 1703 at the first meeting at which Newton presided, was to make the presentation of experiments once again an important feature of meetings from this time until shortly before his death in the spring of 1713.[2]

It is obvious that the very existence of a Curator of Experiments is likely to generate the presentation of experiments at meetings, for Curators had to present experiments with some regularity to justify their existence (and salary). Even Hooke in his later years had been stimulated to some activity whenever the question of payment was raised by the Council, and he was not dependent on his salary alone, having many other interests. Hauksbee, with as far as is known no other post, was presumably far more dependent on his salary than Hooke had been, although he was still active as a lecturer and instrument maker;

Presidents

1703–27	Sir Isaac Newton
1727	Sir Hans Sloane

(1)		Secretaries	(2)	
1687–1709	Richard Waller	1693–1713	Hans Sloane
1709–10	John Harris	1713–21	Edmond Halley
1710–14	Richard Waller	1721–7	James Jurin
1714–18	Brook Taylor	1727	William Rutty
1718–47	John Machin		

Curators of Experiment

1704–13	Francis Hauksbee
1714–44	J. T. Desaguliers

Operator: Henry Hunt, d. 1713 and not replaced

Editors of the *Philosophical Transactions*

1695–1713	Sloane
1714–19	Halley
1720–7	Jurin

Figure 6. Officers, 1703–27.

but in any case he was working directly for the President. He was to prove extremely diligent, so diligent in fact that he presented experiments at some fifteen meetings in 1704, twenty in 1705, offering besides one experimental paper in 1704 and five in 1705. Hence his efforts alone would have altered the tone of the meetings. But he was not alone, for now, for the first time in many years, the President himself was a constant attender at meetings and had a direct influence on their proceedings, which for some years was biassed towards experiment in the physical sciences. Brouncker (P.R.S. 1662–77) had been the last President both to perform experiments at meetings and to report on experiments performed elsewhere at the Society's direction. Wren (P.R.S. 1680–2) had produced some comment in discussion with reference to experiments performed in the past, and Southwell (P.R.S. 1690–5) had occasionally referred to experimental work (some his own) both before and after his term of office, but other Presidents had not been natural philosophers at all. Now Newton was not only to perform several (admittedly simple) optical experiments, two in 1704 and in 1708, but, which is more important, he clearly suggested the subjects of very many of the experiments performed by Hauksbee. These are those well-known experiments which he incorporated into the Quaeries of the post-1704 editions of *Opticks*, experiments from which he was able to draw significant conclusions.

Hauksbee began his presentation of experiments at meetings with a particularly spectacular demonstration of the by then well-known phenomenon of the flash of light produced when a barometer was shaken. Within a few months, although not formally appointed as Curator, he was being paid for the experiments which he brought in regularly and steadily and by the summer of 1704 he had agreed to provide experiments as a firm commitment; for these he was regularly paid, an arrangement which continued unchanged after his election as a Fellow in 1705. He began by demonstrating his vacuum pump and showing a series of experiments *in vacuo*, many of which had their origins in the work of Boyle, for whom, it is conjectured, he had worked in the 1680s.[3] These included (2 February 1704) experiments on the boiling of fluids *in vacuo*, the temperature at which boiling began being lower the greater the vacuum (a phenomenon referred to by Newton in Quaery 11 of *Opticks*), and (12 July) the firing of gunpowder *in vacuo* (which Newton cited in Quaery 10). He also (21 June) read an experimental paper on the condensation and rarefaction of air with tables echoing those of Boyle, a paper on which Newton commented favourably, and showed several examples of the production of light by mixing fluids, a well-known phenomenon performed by Slare as early as 1685, and not original with him. In the next year, 1705, Hauksbee concentrated on the generation of light by friction *in vacuo*, initially deriving from work by Boyle but now given a new direction and new interpretation. First he showed several experiments on 'the mercurial phosphorus' to ascertain whether agitated mercury shone more or less brightly in vacuum than it did in air; by October and November he was pointing out that to obtain the light the mercury must be in motion and the greater the motion, the brighter the light. This led him to conclude that the light was produced by friction between the glass and the mercury, so he tried to ascertain whether there was 'a phosphorus' in the mercury or whether friction of other substances could produce light; this led him in turn to demon-strate (19 December 1705) that a glass spindle rubbed against a woolen cloth *in vacuo* would shine brightly. Not for another year (6 November 1706) did he try rubbing an evacuated glass globe turned on a spindle, with spectacular results. (Newton attributed this effect to effluvia from the glass.) Hauksbee next tried rubbing a barometer tube with the mercury at rest, and when this succeeded he tried larger tubes without mercury (13 November, 11 December). At this point it was seen that the light spread to nearby objects as by attraction, which at first was thought to be magnetic, but, when Hauksbee found that a slight friction was sufficient, he seems to have realised that the attraction was electric. He pursued these results in 1707, showing (8 January) that the best results were obtained by rubbing the evacuated globe with the bare hand and, more important, that threads in the globe moved as the hand approached. Many of these experiments so impressed Newton that he cited them in the revisions to *Opticks* and later in projected (but rejected) additions to the second edition of *Principia*. During 1706, besides his electrical experiments, Hauksbee showed many experiments on capillarity in the form of the ascent of liquids, especially light oils, between plates of marble or glass, experiments which he repeated in 1711 and which were also utilised by Newton. In 1710 he made and described publicly experiments on falling balls in St Pauls, and in the next couple of years some experiments on magnetism suggested by Newton. Thus, not only was he an indefatigable Curator of Experiments, but a highly influential one, whose work is worth examining in detail.

Whether these experiments were all conceived by Hauksbee and influenced Newton, or whether they were all directly inspired and conceived by Newton is impossible to say. But it is clear that the cooperation between the experimenter and the natural philosopher, between Curator and President, was immensely influential and benefitted both in the years from 1704 to 1713. As had been the case with previous Curators, Hauksbee performed experiments most frequently in his earlier years, years which coincided with Newton's work on his first revision of *Opticks*. After this his energies did not slacken, but, instead of performing all his experiments at meetings, he turned to reading papers which contained accounts of the many experiments he had performed elsewhere, some directly for Newton. He then only performed at meetings those which were most suitable for public demonstration, normally at some eight or ten meetings a year (he also read four or five papers describing experiments performed elsewhere during the course of each year). Exceptionally, in 1711, when he was particularly busy working for Newton, on experiments which the latter used in the revised second edition of *Principia*, and in 1712, when there were many visitors to the meetings to entertain, he performed many more public experiments.

Hauksbee's experiments were, of course, all physical in nature; it is worth considering the contributions of others to ascertain whether Newton guided the Society towards more physical experiment than had been the case in other periods. It has been argued (especially by Heilbron in his detailed description and analysis of the rôle of physical science in the Society between 1703 and 1727)[4] that it was experimental physical science which had languished during the pre-Newtonian period, while clearly it flourished greatly under Newton's rule as President. To this end Heilbron considered not only the record of the minutes but also the contents of the *Philosophical Transactions*, obviously a rather different thing, since its editors were free to print what they chose, including material not read at meetings. The *Philosophical Transactions* represent the point of view of the editor, not that of the Society as a whole, and certainly are not an accurate reflection of Society meetings. (As already noted, when Halley was editor he never published anything by Leeuwenhoek.)

It is true that, as has been shown above, the decade before Newton's Presidency saw a real decline in experimental interest, with more interest in subjects, even if often connected with the physical sciences, which were either theoretical or did not lend themselves to the kind of public demonstration akin to that of 'hard-line' physics. Natural history, medical curiosities and case histories, geology – these were, or might be, empirical subjects lending themselves to display or exhibition, although clearly not to any form of experiment, strictly speaking. The same is true of the presentation of models, instruments and drawings, which in seventeenth- and early eighteenth-century terminology might be termed experiment, although moderns would not so regard it. Certainly much of his display was instructive, and certainly it did provide an empirical basis for 'respectable' subjects in years to come, just as much as did records of air temperature and barometric pressure, while it must be remembered that the Society's aim was to improve *all* natural philosophy, not merely its physical branches, and that it would have been contrary to its aims to exclude the non-physical sciences, provided only that they were investigated rationally and empirically. It must also be recalled that it was not yet clear that archaeology, usually then the collection of Roman remains, including coins, medals, objects of general use and inscriptions, was to become a different kind of intellectual enterprise from physics, chemistry, botany or even meteorology; for after all, what is the difference superficially between collecting fossils or minerals or plant and animal specimens and collecting coins? To the seventeenth and eighteenth centuries, all these subjects were equally worthy of systematic investigation. It was to be the nineteenth century, above all in Britain, which was to separate natural science from other branches of learning far more strictly than was the case in Continental countries, and at the same time to begin to exclude medical practice from the Royal Society while retaining medical science. It is therefore ahistorical to censure the Royal Society before Newton's régime for a wider interest than would be allowed now, while, as will appear, it is not the case that the Society under Newton at any time ignored these non-experimental interests, but rather that it continued them even while adding a greater quantity of experiment, much of it physical because Newton's own interests were in the physical aspects of natural philosophy. But it is worth noting that when Newton was, occasionally over these years, not presiding at a meeting, the tone of this meeting was precisely similar to that of those meetings at which he did preside, for, like most of his contemporaries, he was keenly interested in medical reports, case histories and curiosities.[5]

To take an obvious example of non-physical experimentation in this period, consider the case of Leeuwenhoek's letters. These, which, it must be remembered, he had been sending since 1676 with detailed accounts of his microscopical work and his conclusions drawn from it, he continued to send in a regular stream until shortly before his death in 1723. In this period, as earlier, they were always welcomed warmly, translations were prepared when they were written in Dutch, and they were read *in extenso* at meetings, the longer letters often taking two or three meetings to complete, since it was not by any means yet the custom to limit meetings to a single subject or even to two or three subjects, but to have a wide range of topics and many reports at each meeting. Leeuwenhoek's letters not only provided much empirical and experimental substance when read, but usually provoked lively discussion and, frequently, repetition or extension of the experiments he described, as they always had done. There were indeed few years between 1676 and 1723 when one of more letters from Leeuwenhoek were not read amid great interest from those present.[6]

Certainly the presence of Hauksbee and his experiments did not so preempt the time available as to exclude other subjects. In 1704, besides Leeuwenhoek's letters, there was time for the reading of papers on anatomy, on medical case histories, on curiosities, whether mineral

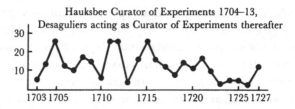

Figure 7. Experiments performed annually at meetings, 1703–27.

(fossils, the bononian stone), medical or natural (including amber-grease). In 1705, the same subjects recur and the natural history content was high, although at the same time Hauksbee performed experiments at some twenty meetings. There was still time for the display of a new hygroscope, for reports of experiments performed by Halley, Hunt and Derham, and for letters and papers on spherical lenses, sound, mathematics, magnetism, and sunspots. In 1706 Hauksbee's contribution was about half that of 1705, with, correspondingly, more papers by others. These were on all possible subjects, experimental, empirical and mathematical, and the same is true of 1707, when there was besides a considerable interest in botany, much of it foreign.[7] The election of Fellows whose interests were not in experimental science certainly had an effect over the years: for example Ralph Thoresby (F.R.S. 1697) contributed papers on Roman remains which were read during meetings in these years, and Newton's friend Stukeley (F.R.S. 1717) might discourse on anatomy or antiquities with equal readiness. The fact that their papers were read at meetings indicates that there was an audience for these subjects. But at the same time experiments continued to be welcome. There were repeated accounts of experiments on phosphorus, there were accounts of analysis of natural and mineral substances (often sent by Geoffroy), and accounts of the action of poisons in 1711, this from both the chemical and medical point of view. In 1712 there were experiments on the generation of insects, and in other

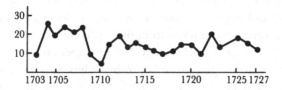

Figure 8. Experimental papers read annually at meetings, 1703–27.

years there was much discussion of the generation of plants. There were, inevitably, accounts of earthquakes and violent storms, at home and abroad, in 1717, 1719 and 1726 auroras were of special interest, and meteors or other unpredictable astronomical phenomena always provoked accounts which were duly read at meetings. Eclipses and comets were, naturally enough, always observed and accounts of them read. Indeed there was a completely general interest still in the whole natural world. Medical science, geology, botany and zoology suffered alike from the obvious fact that possible instrumentation was small and that the microscope could not compete with airpumps, barometers, watches, hygroscopes, lenses, anemometers or electrical machines for interest and display, so that true experimentation was less easy than for the physical sciences, and when performed less suitable for display. But it must be noted that many magnetic experiments did little more than appeal to the same interest in the curious that natural history more obviously provided.

No one subject could absorb the attention of the Society, not even skilful experimentation by Hauksbee. Thus in 1711, when Hauksbee performed as experimenter or paper reader at some twenty meetings, there was still time for a detailed review of French experiments on the effects of poisons on dogs, together with the reading of a letter which recounted experiments to determine quantitatively the loss of fluids by both humans and animals and of a demonstration of the anatomy of the frog as well as for the reading of letters from Leeuwenhoek, chemical accounts of the analysis of ashes and several papers on various subjects by Papin. In 1712 Hauksbee again performed at over twenty meetings, giving time for descriptions of anatomy, magnetic experiments, papers on optics, and other topics.

When Hauksbee died in 1713, followed a few months later by Hunt, the long-time Operator, experiments naturally became fewer, but the tone of the meetings did not greatly alter. In fact, the hiatus in Curators of Experiment lasted only a year, for in 1714 the appearance of J. T. Desaguliers, promptly elected a Fellow, led to renewed experimental demonstrations immediately. And, although never officially Curator of Experiments, Desaguliers acted as one for many years, although the number of experiments he actually performed before the Society became ever fewer over the years before his death in 1744. He was never to be either as assiduous or as original as Hauksbee had been, most of his experiments being derivative, mainly repetitions of those already recorded by Newton or, if novel, performed at Newton's instigation and presumptive invention. Thus in 1714 at Newton's request he repeated Newton's former experiments on heat, while Newton himself both showed and discussed a parallel experiment on the heat to be derived from fermenting bran. In both 1714 and 1715 Desaguliers showed and described (that is, wrote up and read out the account) a number of experiments on light and colours which were intended to confirm the accuracy of Newton's own, earlier experiments, often challenged over the intervening years. In 1714 Desaguliers also showed 'blood globules' in capillary tubes and, with Brook Taylor, exhibited several magnetic experiments, designed, as Taylor's paper read later that year put it, 'to discover the Law of Magnetic Attraction'. During 1714 there was discussion at several meetings, initiated by Newton, about the reaction of aqua fortis and salt of tartar (nitric acid and potassium carbonate), an experiment familiar to many of the members. The cattle plague then raging in England, Holland and France was described at several meetings, there was a paper on the advantages of the use of inoculation against smallpox, with empirical examples, and a letter on the structure of animal fibres. In 1715, besides the experiments already mentioned on light and colours, Desaguliers showed experiments on pneumatics, magnetism, hydrostatics and mechanics, usually several on each of the occasions at which he presented his work. There were in 1715 other experimental contributions from Fellows: Dr A. Stuart (F.R.S. 1714) showed an experiment to demonstrate the existence of air in all animal fluids, there was a description of anatomical preparations, Dr Steigerthal and Martin Folkes both described anatomical experiments performed by themselves privately, and Sloane, no longer Secretary, described his work on the propagation of plants following a Dutch account read at a previous meeting. There were also papers and letters, all empirical, and most more or less experimental, on botany, anatomy, magnetism, and meteorology.

And so it was to continue for the next few years, with Desaguliers consistently either performing or describing experiments on optics, magnetism and mechanics. Before 1720, indeed, the bulk of the experimental content of the meetings was supplied by him, and consequently was physical in nature. Later, there was to be a wider variety of participants, and, perhaps under the growing influence of Sloane, there came to be a greater emphasis on the biological side. But even during Newton's later years as President, biology and medicine were always well represented. In 1716 there were two serious botanical papers by Richard Bradley, one experimental, the other observational. In 1717 Dr Stuart read an experimental paper on the changes of colour to be observed in leaves and flowers, and there were papers on anatomy with new observations. In 1718 there were several papers on surgery, including one by Jurin on the force of the heart; in 1719 he contributed a description of his experiments on the specific gravity of the blood. In both these years letters from Leeuwenhoek commanded time and interest. In 1719 there was a good account of the 'contagious Distemper which raged amongst the Milch Cows in the year 1714' (as noted above, it had been briefly described in its epidemic year). There were several papers on plant and human physiology, anatomy and medical problems, including the case of an anorexic boy by Dr Patrick Blair (F.R.S. 1712), which Sloane received with reasonable, although with modern hindsight perhaps excessive, scepticism. More important, Stephen Hales contributed a paper describing the effect of the sun's warmth on making the sap rise in trees and was almost immediately elected F.R.S.; the phenomenon described had been studied by several Fellows nearly fifty years before, but never, as Hales was to do, quantitatively. And Desaguliers contributed a paper on animal respiration.

These biological papers were, more often than not, based upon experiment and always on empirical evidence, but they did not in general lend themselves to experimental demonstration in the Society's rooms. Physical experiment was easier to show. So Jurin in 1718 showed experiments intended to discover whether air entered into the pores of water as well as describing measurements of the flow of water through tapering tubes, while in 1719, in 'answer' as he said to experiments performed earlier by Hauksbee and Taylor (in 1712), he described experiments on the flow of mercury through capillary tubes. In 1709 Sir Godfrey Copley (F.R.S. 1696) had left a sum of money to provide experiments at meetings; now in both 1718 and 1719 Desaguliers produced experiments which he stated were 'to serve for the Annual Experiment Appointed to be made by the will of Sir Godfrey Copley':[8] in the first year a repetition of the by then well-known feather and guinea experiment demonstrated the difference between free fall in air and in a receiver evacuated with an airpump, and in the second year he demonstrated a working model of an improved Savery steam engine. In this latter year he also showed examples of electrifying glass by friction, of firing gunpowder *in vacuo*, of rarefying mercury, of the resistance of air to falling bodies (also describing experiments on falling bodies inside Wren's St Paul's), and described the old experiments of Villette with a burning glass.[9] This last prompted both Newton and Sloane to describe burning glass experiments made by them in the past. There were, as usual, several astronomical and meteorological accounts, while (in 1719) Derham's description of a ring around the moon led Newton to give an impromptu account of sunlight, and in the same year Halley read a paper on mock suns, and there were several descriptions of the aurora borealis. There were also in 1719 several experimental papers sent in from outside the Society, including one by Fahrenheit on the rarefaction and specific gravities of fluids. Besides all this, there were papers on fossils, accounts of distant lands, Stukeley's important study of Stonehenge, Conduit's account of Roman towns in Spain, and so on. These were all very similar to papers and letters read in the past and all, in the spirit of the times, empirical, although obviously far from being experimental. Thus, without Desaguliers, physical experiment would have been less frequent than biological experiment, and it was only his presence which maintained the balance in its favour.

In 1720 Desaguliers was still very active, reading or presenting experiments at about half the meetings. These varied from demonstration of the barometer flash (often shown earlier) and comments on an important electrical paper written by Stephen Gray, a newcomer not yet F.R.S., to respiration experiments, specific gravity measurements and more on the flow of water in pipes and through holes, to the showing of new machines and instruments, among them a barometer, a ship's log and a portable boat. Halley presented theoretical papers on astronomy and on mathematical optics. Experimental papers included one by Whiston (never F.R.S.) on the dipping needle and an account of Dorset mineral water by 'Mr Godfrey Chymist'.[10] Once again, biological and medical papers preponderated, although not greatly, over papers on physical subjects. There were the usual medical and anatomical case histories, by no means always by physicians. Thus, among the latter, was an account by Newton of a death by lightning which had occurred about thirty years before in Grantham which led to a considerable discussion about the functions of the heart and lungs, while Halley described the use of quarantine in the Mediterranean world to control the spread of plague. A thorough account of the King's Evil (scrofula) and the use of the royal touch to cure it was given in critical terms by William Beckett (F.R.S. 1718). There were a number of botanical papers: on the medicinal virtues of plants by Dr Patrick Blair (F.R.S. 1712), a letter from Jussieu in Paris, and an experimental account of grafting with a discussion 'of the motion of sap in plants and trees' as it affected the process by a gardener, one Thomas Fairchild, who had previously shown his experiments on grafting to the Society; now his paper was thought interesting and important enough for publication by Halley in the *Philosophical Transactions*. And there were the usual natural history papers on oysters, bees, and so on.

In the next year, 1721, Desaguliers was less active, but in compensation the biologico-medical interest continued strongly. the lithotomist Cheselden (F.R.S. 1711) reported on his practice as did Douglas on his; there were letters on abnormal pregnancies, observations on the epidemic of plague at Constantinople, anatomical accounts (including one on the dissection of an elephant), an account of a smallpox epidemic in New England, and the use of inoculation there. These accounts all seem strangely remote from the natural philosophy of Newton, but he clearly did not so regard them, for to him, as to the Society generally, medicine was a topic of interest and relevance, certainly not seen as being outside the Society's concern nor as differing in quality and method from botany or physiology. In botany during 1721 papers varied from experimental (as in a paper examining the 'force of germination') to natural history (a description of a curious fungus and another of fossil trees) to the merely curious (an African wood displayed by Desaguliers, but as this gave a red dye it may have been considered less as a curiosity than as a potentially useful substance). Inventions were shown: John Hadley (F.R.S. 1717) produced his reflecting telescope, Desaguliers tested a pump invented by Joshua Hoskins, and there were shown models of a bridge, of a kiln for drying malt, of a water raising machine devised by 'Mr Harding' and of Halley's diving bell; a glassmaker showed examples of stained glass, there was a paper by Du Quet on an invention of revolving oars for war galleys, and Halley described the discovery by a watchmaker named Williamson that a pendulum clock orientated east–west and one orientated north–south diverged significantly in time, although started showing the same time. There were a number of astronomical papers, mostly observational, by Halley, Bradley and others, but few physical experiments. True, Desaguliers described what he called experiments on the amount of water in the New River; he showed genuine experiments on the resistance of fluids by using a pendulum with a bob in for form of a gold sphere which vibrated in a trough filled first with water and then with mercury (30 March, 20 April), experiments which later produced comments from Newton; he showed an apparatus with which to demonstrate the impossibility of perpetual motion (a kind of balance) but did not make the experiments; and he twice performed simple optical experiments on opacity. But this was far less than he had shown in previous years, and true experiments whether by him or others were very few indeed. A chemical display of a 'new ignitable phosphorus' by a chemist named Brown stands out as being exceptional.[11]

Clearly by the early 1720s it was not merely that experiments in general were fewer than in previous decades but that physics was less strongly represented while medico-biological science was very strong, possibly dominant. It has often been suggested that it was Sloane who produced this effect.[12] But not only was Sloane not especially dominant in this period, there were others besides him who leaned in this direction while being more active in the physical sciences than he was or could be. Both Martin Folkes, at this time a Vice-President, and James Jurin, Secretary and editor of the *Philosophical Transactions*, were ardent New-tonians and competent mathematicians. But Folkes was also deeply interested in antiquarianism, and Jurin, although he presented and published papers in experimental physics, was a physician best known for his work on quantitative experimental physiology. Newton was no longer raising questions which Desaguliers could solve by experiment, and Desaguliers, without Newton's lead, was clearly reluctant to perform experiments unless he was sure of receiving money for them from the Copley fund (or presumably from his audience at private lectures). Newton had ceased to plan fundamental revisions to either *Opticks* or the *Principia* of the sort requiring experimental confirmation. He was now a very old man (eighty in 1722), and although he attended the Society's meetings regularly, met visitors and clearly had all his wits about him, he appears to have left much of the running of Society affairs to his disciples, all relatively young. The Society was slowly taking its eighteenth-century form, most obviously by the nearly complete replacement of experiment and spontaneous discussion by increasingly formal and fewer, perhaps because longer, papers. The fairly formal presence of 'strangers', visitors introduced by individual Fellows, probably also reduced spontaneity of discussion.

So the year 1722 opened with the reading of papers both physical and biological. These included a description of a new barometer, a botanical paper from New England, a medical paper from France (experimental, this, since it was an examination of the bile of plague patients), an account of the aurora borealis recently seen in Devon, and Desaguliers' account of his experiments on the descent of fluids. This mixture of subjects may seem to moderns so diverse as to indicate a lack of serious purpose, but it was perfectly in the spirit of the Royal Society of sixty years earlier except that there seemed little inclination to show experiments to the assembled Fellows; perhaps the criticism should rather be that in sixty years very little had changed. The remainder of 1722 was not to be very different. There were accounts of instruments which aroused interest, among them Colonel Molesworth's new sandglass, described by Newton and then displayed at a meeting so that all could judge its performance (1 and 8 March); it was tested, perhaps by Desaguliers, for three months later there was a report, on which Newton commented, indicating that the rate of flow was not truly constant. There were astronomical reports and comments thereon, and meteorological ones (of auoras, parhelias, a strange sunset, rainbows and hurricanes); there was an earthquake and volcanic eruption off the Madeira Islands which produced accounts; there was plague at Marseille, which produced several medical reports of dissection findings; and the continuing smallpox epidemic prompted much interest in the results of inoculation. Only a few experiments were displayed by Desaguliers: one intended to prove the falsity of Leibniz's principle for explaining the rise and fall of a barometer (17 May) and (6 December) a repetition of one of Newton's experiments in *Opticks* to refute an Italian doubt that differently coloured rays of light did truly possess different degrees of refrangibility (cf. p. 146 below). But when Jurin (also on 6 December) sought to 'vindicate' his theorems about the force of water running out of a cistern, neither he nor Desaguliers who read a paper on the same problem a fortnight later attempted to show any experiments.

In 1723, although meetings seem to have continued to draw a good attendance, and Newton usually presided, there were no real experiments performed during the whole of the year. There were, it is true, several experimental accounts: a French description of a new fire extinguisher, Brook Taylor's account of an experiment to prove that the expansion of a fluid is truly a measure of heat (14 March), 'Mr Cay's' letter on experiments on the specific gravity of water impregnated with salt,[13] and on the specific gravity of metals (20 June), and a letter from Mr Brown describing the manufacture of Epsom salts and his comparison of the chemical with the natural product (4 July), analyses by an Oxford chemist, among the more interesting examples. Leeuwenhoek's last letter (he died at the end of August) was read in June. There were many papers on smallpox and on inoculation, including a statistical paper by Jurin analysing the value of inoculation and a microscopical examination of smallpox exudations by Desaguliers, who otherwise contributed little except (31 January) a 'draught' of an aurora which he had seen the previous month, and which was also described by others. There were several meteorological accounts, and Sloane moved that a rain gauge be prepared 'for the Society's house' after a Mr Horsley had spoken on the preparation of a weather register.[14] There were also several astronomical, mathematical and magnetical papers. Nor was 1724 to show any changes, except that, according to the minutes, Desaguliers was hardly active at all until the end of the year. His contributions were limited to a paper defending Newton's view of the shape of the earth against the French view, a hotly debated topic to which Desaguliers added some useful experimental demonstrations, and the description of a barometer to be used as an altimeter.[15]

In 1727 the experimental content of the meetings was raised considerably by the contributions of Stephen Hales. His paper, read under the title 'A treatise concerning the power of vegetation', to be published three years later as *Vegetable Staticks*, was read out at a number of meetings; few natural philosophers were more committed to Newtonian experimental work than was Hales. There continued to be many anatomical papers (one of which at least, by Douglas, could count as experimental),[16] as well as several case histories and much more on inoculation. Desaguliers continued to read on the figure of the earth, including a paper, said to support Newton's view, describing cohesion experiments on lead,[17] and also described a machine of his invention which was intended to demonstrate the correctness of Newton's theory.

In this form, as the minutes reveal, the meetings continued for some years to come, the death of Newton in March 1727 and the subsequent election of Sloane as President producing little change. Desaguliers continued, until his death in 1744, to perform one or occasionally two experiments in each year. He presumably hoped to receive the money from Copley's donation for these, as he did in 1729, 1734, 1736 and 1741 together with the Copley Medal, instituted in 1736 at the suggestion of Folkes. But Desaguliers did not receive the Copley award in every year before 1744: although he performed an experiment on 11 February 1731 the award went to Stephen Gray for his experiments on electricity, as it did in the next year as well.[18] In 1735 and 1736 Gray's experimental papers were read at length at several meetings (they were also published in the *Philosophical Transactions*). Few of the Copley awards at this time (in 1738 and 1743 no award was made) involved the direct presentation of experiments, description alone sufficing, while after 1756 the Council decided that the award might be made for papers published at any time and not necessarily in the year of the award.[19]

The actual performance of experiment at meetings thus became increasingly rare. The few by Desaguliers have already been noted. Some were occasionally performed by others. The chemist Mr Godfrey presented an experiment in 1732, and in 1736 some experiments devised by Gray were posthumously performed as described in his papers.[20] Papers containing accounts of experiments continued to be presented, along with miscellaneous empirical papers on many familiar topics. Instruments, machines, 'specimens of a sort of fine white ware' (10 February 1743), and so on, continued to be displayed, together with rather fewer random natural curiosities. What is notable is the increasing formality of the 1730s and 1740s, with discussion growing ever less, accompanied by occasional complaints at 'unruliness', which indicates that discussion was limited to private discussion between individuals.[21] Not even the reading of the minutes of the preceding meeting seems normally to have provoked discussion, or if it did it is not recorded. It is probable that not all letters and papers sent in were now read; this was the period when the Council began to show more concern for the *Philosophical Transactions* and the standard of its contents. By 1743 each paper printed therein contained a notice of the date on which it had been read to the Society. There was now no room for informal presentation; the papers might be of a higher standard, but there was less room for the presentation of experiment or of discussion. There were to be no more Curators of Experiment after Desaguliers. Nor were there to be persons like Gray and Godfrey who, not being Fellows, (Gray was elected only in 1732) were yet encouraged to present their experiments in persons to the Society. Slowly it became the custom, and then the rule, that papers were read by a Secretary. At the same time the spoken and printed word replaced ocular demonstration; those experiments which were described were accepted at face value provided always that their description and the conclusions drawn from them seemed reliable. It worked well on the whole, but might, obviously, prejudice the acceptance of really novel discoveries, as it clearly sometimes did. Well might Marshall Hall complain at the rejection of his experimentally based conclusions without his being allowed to show his critics the experiments themselves. He could have pointed out that the Society's proud motto of *Nullius in verba* had little or no force when the Fellows no longer wished to see for themselves but were prepared to rely upon interpretation of the written word.

~ 8 ~

The communication of experiment
1703 ~ 1727

With the inauguration of Newton's Presidency, much was to change within the Society and with the perception of it by outsiders. Most strikingly, it is for this period impossible to separate the office from the office-holder, the Royal Society from its President, and Newton the natural philosopher from Sir Isaac Newton P.R.S. The two rapidly became one in most eyes, so that the Society received the credit for Newton's fame and influence, while he in turn took on the attributes of the Royal Society.

Relations between the Society and individuals at home and abroad were also coloured by the fact of Newton's Presidency. Sloane remained an active Secretary for ten years, corresponding as before with, particularly, natural historians, medical men, and his French friends (he joined Newton as an associé étranger of the Académie Royale des Sciences in 1709); after that as a Vice-President he presided at meetings when Newton was not present and of course continued much of his correspondence and influence, although less publicly. In 1713 Halley became Secretary so that now official correspondence took on a more astronomical bent. The publication of *Opticks* and the growing number of adherents to the Newtonian natural philosophy meant an enormous growth in knowledge and understanding of the Royal Society's empirical programme, now ineluctably merged with the dominating figure of its President, even though the cult of Newtonianism associated with the Enlightenment lay still in the future. Learned visitors were eager to view the Royal Society and its President, one of the essential 'sights' of London. And the great controversies over mathematical priority and the correct approach to natural philosophy which raged between Newton and Leibniz[1] necessarily publicised both Newton and the Royal Society, while their disciples corresponded extensively during the affair. English Newtonians were ever eager to inject the Royal Society's empirical philosophy into their propaganda and did so whenever possible.

The publication of *Opticks* (1704), which followed almost immediately upon Newton's election as P.R.S., was of enormous importance in the publicising of the Royal Society's experimental programme. The original English edition was, unusually, distributed abroad to a certain extent: in particular, Geoffroy possessed a copy by 1705, in which year he began to read a summary of it to sessions of the Académie Royale des Sciences, readings which continued in 1706.[2] The *Acta Eruditorum* of Leipzig could only review the mathematical papers which were in Latin in 1705, but as soon as the Latin edition was available the *Acta* reviewed it at length, as did the *Journal des Sçavans*. The experimental physics practised at the Royal Society was, as already noted (Chapter 7), utilised and enshrined in successive editions of *Opticks* in both English and Latin by means of the increasingly numerous Quaeries, which cited the experiments which Hauksbee had, both privately and at Royal Society meetings, performed at Newton's instigation. Moreover, Hauksbee's own accounts were available in English first by publication in the *Philosophical Transactions* and then in his own *Physico-Mechanical Experiments on various Subjects* of 1709, which was published in Italian in 1716, and so made available to many non-English readers. Further, Newton's numerous disciples like John Keill, John Freind and many others, as well as writers on natural theology, were publishing their own versions of 'Newtonian' natural philosophy in English and in Latin, variously concerned with physics, astronomy, chemistry and medicine, all now conceived in a Newtonian mode. A most remarkable example was the work of Samuel Clarke (F.R.S. 1728). He produced first a Latin and then an English translation (1697 and 1723, respectively) of Jacques Rohault's immensely popular *Traité de physique* of 1671, with copious Newtonian footnotes contradicting the Cartesian text, a means by which English undergraduates were, in the eighteenth century, introduced simultaneously to natural philosophy and Newtonianism. Even botany could be Newtonian, as the work of Stephen Hales brilliantly showed. His *Vegetable Staticks* (1727, first partly read to the Royal Society's meetings in 1719) was avowedly Newtonian; it is not surprising that when it was published in French in 1735, the preface of Buffon (who translated it) was as much concerned with the importance of the English experimental method, which he praised highly, as it was with the work of Hales.

Peculiarly important was the work of the Dutch natural philosopher 'sGravesande. He came to England on a diplomatic mission in 1715 and soon was in touch with the Royal Society, to which he was quickly elected. Even before this, he attended a meeting (24 March) at which Desaguliers had shown a number of Newtonian experiments to an audience of several 'strangers'. (The rule forbidding visitors had been first relaxed, then repealed (1698/9); their presence, now quite customary, was once again an important means for the spread of the Society's experimental ideas.) After his appointment as Professor of Mathematics and Astronomy at Leiden, 'sGravesande widely publicised the Newtonian ideas with which he had been inculcated in England, filling his lectures with material drawn from both the *Principia* and the *Opticks* and demonstrating scientific principles by means of experiment on a regular basis. His anglophile, Newtonian approach was even more widely diffused by the publication in 1720–1 of his *Physices elementa mathematica, experimentis confirmata*, a work which went through many editions, was almost immediately translated into English, and some twenty years later into French. In its Latin dress it was read by far more people than could ever aspire to read the *Principia*, and probably by far more Continental natural philosophers than read even the Latin translations of *Opticks*. It aimed to publicise Newtonian natural philosophy and the use of experiments to demonstrate the principles of theoretical science; in so doing it also publicised the methods and aims of the Royal Society. So too there were published several influential works on chemistry which stressed the importance of the Newtonian or Royal Society empirical methods. These drew both on the chemical Quaeries to *Opticks* and on the work of Robert Boyle, which seemed close enough to Newton's in spirit to share in the demonstration of the new approach. Thus for example Peter Shaw drew impartially on experiments derived from both Boyle and Newton in his *Chemical Lectures* of 1734, as he did in translating Boerhaave's *Institutiones chemiae* (1724) as *A New Method of Chemistry* in 1727.

The *Philosophical Transactions* in the early eighteenth century also benefited greatly from Newton's influence and his direction of meetings.[3] Hauksbee's experimental discourses were all published by Sloane during his editorship (through 1713) as were the earliest papers by Desaguliers. Halley took over the editorship in 1714 for six years, publishing the work of Desaguliers regularly and much physical, although little biological, science. The more catholic Jurin was editor from 1720 to 1727, to be succeeded for one year only by Dr William Rutty (F.R.S. 1720) who produced the volume for 1728; he was followed by Dr Cromwell Mortimer (F.R.S. 1728) physician, chemist and antiquary, who served as secretary from 1730 to 1752. During the second half of the eighteenth century the secretaries who edited the *Philosophical Transactions* were of a literary bent: Thomas Birch, editor and compiler, who served from 1752 to 1765 was followed by Dr Matthew Maty, Secretary 1765 to 1777 and Joseph Planta, Secretary 1776 to 1804, both librarians at the British Museum. That the editors were not natural philosophers was no longer important to the content, for in 1752 the journal became what the learned world had always taken it to be, no longer a private venture of the Secretary but the official publication of the Royal Society, whose Officers and Council consti-tuted a Committee of Papers to select those suitable for publication. Publication continued without break. Editors noted now when papers were read, and gradually all papers published had been submitted expressly for reading and, if possible, publication. Thus it was that readers of the *Philosophical Transactions* had a clear understanding of what occurred at meetings, for the reading of papers was now the principal business. There was virtually no discussion. Once a year there was printed a list of the 'presents' (chiefly books) presented at previous meetings, so that only accounts of elections and the names of visitors and their hosts were omitted from the printed account.

Many (like Geoffroy) now kept in touch with what was happening in the Society throughout the century by diligently reading the *Transactions*. Geoffrey was particularly interested in Hauksbee's work, requesting more details from Sloane about the experiments with the airpump and 'particulierement celles de la refraction' (a comparison of refraction in air and the vacuum), which the French Academicians had tried unsuccessfully to ascertain in 1700.[4] Geoffroy remained a great admirer of Hauksbee, and when, in 1715, his brother, with Pierre Rémond de Montmort and the Chevalier de Louville, visited England, he provided letters of introduction to Sloane with the particular request that they might see some of Hauksbee's experiments – he did not know that Hauksbee had died two years before this. The visitors were shown experiments especially designed to interest them, including one by Desaguliers concerning the solubility of salts and the effects of solution on them which he had derived from the elder Geoffroy. They also wished to see Newton's experiments on colours, and these Desaguliers indeed repeated during the course of 1715. The visitors were all predisposed to admire the Royal Society (as indeed they did): Montmort had earlier displayed his admiration of Newton by sending him one of his books. They were all delighted to be elected Fellows almost immediately, increasingly the practice with foreign visitors and benefactors under Newton's Presidency. Both the Geoffroys and Louville were already members of the Académie Royale des Sciences, while Montmort was to be elected to it the next year, so the honour they received from the Royal Society strengthened the bonds between the two institutions. Louville had evidently been in touch with Newton earlier, for when in 1714 Fontenelle, Secrétaire perpétuel of the Académie since 1699, wrote to thank Newton for 'un Receuil de different piéces de vous', he noted that they had been transmitted by Louville.[5] Montmort was to become a correspondent of Newton's after his return to France. Thus, long before the high days of the French Enlightenment, Newton and the experimental philosophy of the Royal Society were known and apparently esteemed within the Académie Royale des Sciences, if only by a minority. And concomitantly, the Royal Society's experimental work and experimental methods were publicised to a French audience thus early.

There is somewhat less direct evidence for Italian reception of Royal Society news during the period of Newton's Presidency.[6] Correspondence with Italy in the early eighteenth century was never on the scale of that in the 1670s and 1680s. It was chiefly conducted through Englishmen resident in Italy and through English travellers there, while there were also a surprising number of Italian travellers to England. From Florence, Dr Henry Newton (no relation to Isaac Newton) wrote regularly to the Society between 1709 (when he was elected F.R.S.) and 1714, occasionally directly to his illustrious namesake,[7] and Sir Thomas Dereham, Bt. (F.R.S. 1720) did the same between 1720 and 1728. (In the same way, towards the end of the century, Sir William Hamilton (F.R.S. 1766) wrote from Naples, but in his case mainly with his own contributions on earthquakes and antiquities.) Among English travellers briefly in Italy was Alexander Cunningham, an historical writer, never F.R.S., who had met Newton in 1703 and wrote directly to him in February and April of 1716 from Venice, where he met many important Italian natural philosophers, including Bianchini, who had been in England, and Poleni (F.R.S. 1710), to whom he gave copies of *Commercium Epistolicum*; he reported his acquaintances as being favourable to the English side in the Leibniz–Newton dispute.[8] Among Italian travellers should be mentioned Bianchini, elected F.R.S. at the time of his visit in 1713, astronomer and mathematician, who had four papers published in the *Philosophical Transactions* over the next fifteen years (1713, 1724, 1725 and 1729); Marsigli (F.R.S. 1692) who came to England in 1722, when Newton received him most cordially and praised his achievements at the first meeting he attended; and the Abbé Conti (F.R.S. 1715) when first in England. (Conti played an active role in the calculus dispute, mainly on the Newtonian side, writing widely to France and Italy; unfortunately he ultimately lost Newton's confidence.)

The best means of communication was undoubtedly through Newton's own books.[9] Thus Guido Grandi (F.R.S. 1709), Professor at Pisa, received a copy of *Opticks* in return for his own mathematical works which he sent to Newton in 1703. *Opticks* was quickly and widely disseminated: in Rome and Naples through Henry Newton, who reported that copies were welcomed eagerly. When Bianchini met Newton in 1713 he gave him the names of a number of distinguished Italian mathematicians and natural philosophers, and to four of these Newton sent copies of the second edition of the *Principia* as it came from the press, while to Bianchini (who had repeated successfully a number of Newton's optical experiments) he gave a copy of *Opticks*. One of the four recipients of the *Principia*, Celestino Galiani (F.R.S. 1735), had read *Opticks* in Naples in 1715 after learning of it through correspondence with 'sGravesande, who had sent him copies of the Newtonian treatises of Whiston and Keill. The chief period of the dissemination of Newtonian ideas, however, came after Newton's death, partly directly, partly through the works of English Newtonians like Derham, whose *Astro-theology* of 1714 was published in Italian in 1728, and George Cheyne, whose *Philosophical Principles of Natural Religion* of 1705 was published in Italian in 1729; it is paradoxical that works on natural religion should have been permitted to pass all censorship while works on natural philosophy were still restricted. Later still came the diffusion of Newtonian ideas through the work of Algarotti (F.R.S. 1736), though his famous *Il Newtonianismo per le dame* of 1737 was directed at the virtuoso, not the natural philosopher, and the works of Voltaire, both these men having also travelled in England. Communications between Italy and England were good enough so that twenty-eight Italians were elected to the Society during Newton's Presidency, eight of them travellers to England.

Not surprisingly, during the prolonged dispute between Newton and

Leibniz few Germans were elected F.R.S.; these few were usually medical men. (Paradoxically, both Leibniz's Swiss disciples, Johann and Nicholas Bernoulli, were elected in 1712 and 1713, respectively.) But in fact Leibniz approved of Newton's 'work on colours', and there was a favourable review of *Opticks* in the *Acta Eruditorum* for February 1706, which summarised the arguments and reported with approval on Newton's use of experiment. Other factors were of course at work to deter the Society from electing Germans, like the German dedication in this period to a more mystical chemistry than the Royal Society could tolerate. (When Peter Shaw (F.R.S. 1751) translated Stahl, he omitted all Stahl's arguments based on non-material substance.) There were few German scientific societies active before 1730 with which the Royal Society could maintain contact: Sturm's Collegium Curiosorum had ceased to be active when its founder died in 1703, and the Berlin Academy, founded in 1700, was too much Leibniz's foundation to appeal to Newton's Royal Society. Cultural contact in natural philosophy failed to follow the crown after George I's succession to the throne in 1715, and German natural philosophers on the whole were later to look to France rather than to England for method and inspiration.

The great days of the Royal Society's devotion to experimental learning and the performance of experiment might lie in the past, but the seventeenth-century legacy of its activities and of those of its members was still highly esteemed in the eighteenth century. The Royal Society remained an important part of the world of learning, playing an active rôle by virtue of the accomplishments of its members and its own reputation for the nourishing and promoting of empirical natural philosophy. Gradually the Royal Society's cherished experimental method was generally accepted. Everywhere in Europe English empiricism was widely known, although now subsumed under the broad term 'Newtonianism'. Newton's legacy, which he had derived originally from the founders of the Royal Society, was to find its place in the European culture of the Enlightenment from Russia to Italy. But when Newton died in 1727 all that lay in the future, although one of its most famous promulgators, Voltaire, is said to have arrived in London, symbolically, on the very day of Newton's funeral. Sadly, Newton left no such decided legacy to the Royal Society of which he had been President for twenty-four years, and the experimental content of the Royal Society's meetings, which had already dwindled under his aegis, became virtually lost in formal papers and letters.

~ 9 ~

The view of the world: friend and foe

As should be all too obvious by now, the Fellows' view of the Society to which they were proud to belong was of an organisation which flourished in proportion as its meetings were devoted to experiment and empiricism. This they saw as their primary aim, this they endeavoured to put into practice, this they tried to promulgate and publicise, it was to this that they turned when the Society appeared to languish and need reform. They were perhaps almost too successful. For the fluctuating reputation of the Royal Society over the first seven decades of its existence was very largely in proportion to the quality of such work as perceived by the public: as performed and discussed at meetings, as reported in the *Philosophical Transactions*, and as revealed in the publications of individual Fellows. The intellectual world on the whole agreed with the Fellows that empiricism lay at the heart of the Society's activities, although not all agreed on the value of such work.

It is difficult to find impartial observers, necessarily outside those more or less closely in touch with the Society, not least because, in spite of occasional proposals to the contrary, in the seventeenth and eighteenth centuries the Royal Society was never an exclusive body, numbers were not limited and so those who sympathetically praised the Society were generally elected forthwith, whether active natural philosophers, virtuosi, physicians, natural historians, antiquaries or noblemen.[1] Intellectuals not so honoured were apt to view the Society with disfavour, no doubt a prime reason for their not being elected. But in all comments, pro or con, it was the experimental basis of the Society which first caught the eye of outsiders and caused them either to praise the Society, and even wish for election to it, or to castigate it for what was taken to be an erroneous view of the proper way to study nature or even, sometimes, of the proper employment of the human intellect, which should devote itself to the higher spheres of thought like religion and theology rather to the lower regions of the mutable, fallible subjects of nature and man.

It is worthwhile briefly to consider the favourable views taken by those not members, or not yet members, who approved of the Society's activities. Much of these activities were easily comprehensible to those with no desire to participate in them: as Francis Bacon had noted, experimental and empirical investigations were comprehensible to many who could hardly, if at all, understand the theoretical concepts of mathematical or complex natural philosophy. Such men could collect material by observation and report, even when they did not know how to utilise their material; so too they could admire the Society's contributions to natural history and hope to add a little to them. An example of a phenomenon to be often repeated in succeeding decades is the work of Joshua Childrey (1623–70), a country clergyman who in 1660 published *Britannia Baconica*, a collection of miscellaneous facts of natural and medical history, inspired as its title suggests by Bacon's *Sylva sylvarum*. In 1669 and 1670 he wrote numerous letters to the Society filled with similar facts; these were read at meetings from time to time and Oldenburg published extracts in the *Philosophical Transactions*. He was clearly a virtuoso not very different from many in the Society, but he was never proposed for election, perhaps because he was not personally known.[2] A keener, but less productive, virtuoso was Samuel Pepys, but he had many friends among the Fellows and was elected F.R.S. in 1667 (by which time he was a not unimportant member of the civil service). When in 1665 he heard from his friends accounts of experiments performed at the Society's meetings he listened enthralled, and then proceeded to buy books on experimental philosophy: Hooke's *Micrographia*, which he described as 'pretty', and 'the most ingenious book that I ever read in my life' and 'Mr Boyle's book of Colours which is so Chymical that I can understand but little of it, but understand enough to see that he is a most excellent man'. He faithfully read these (to him) difficult works, but it was the experiments which interested him, partly of course because he could understand them.[3] Like others, he much enjoyed seeing experiments performed at the meetings he attended after his election, finding them greatly entertaining. Hence, at a dinner party he gave during his Presidency, he encouraged Slare to show an experiment in which the mixing of two liquids produced particles which gleamed like fire or the stars, to the party's great pleasure.[4] Just so some of the experiments related in Boyle's 'book of Colours' had been shown by the author to ladies for their entertainment when they visited him, as he there stated. Those Fellows who were not practising natural philosophers nearly always described meetings in terms of the experimental entertainment. Abraham Hill, an original Fellow, but an administrator rather than a true virtuoso, described various meetings held in 1663 to a fellow-member almost entirely in terms of the experiments performed.[5] This continued to be the case in later decades whenever men with only a mild taste for natural philosophy attended meetings, only by then they were forced to record accounts of experiments given in formal papers rather than describing experiments they had seen performed.[6]

In large part through the work of the various Secretaries, (especially Oldenburg) on the one hand, and the publicly proclaimed views of the more active Fellows on the other, most foreigners accepted the Society's own view of itself and saw it in terms of experiments performed and empiricism espoused, and this from the earliest days and in all countries (though with varying degrees of approbation). Their sources for this knowledge have been discussed above: the works of prominent Fellows, most readily those published in Latin, correspondence, the *Philosophical Transactions* (even in the troubled period of the last decades of the seventeenth century, when publication was spasmodic), sometimes available through the reprinting of excerpts from it in French, German or Italian periodicals, all these convinced foreigners that English natural philosophy excelled above all things in empiricism, while on the whole often being deficient in theoretical and, especially, in metaphysical foundations. The correspondence, mainly conducted by successive Secretaries, although often irregular in the later seventeenth century, was clearly always valued. And foreign travellers continued to be proud of the opportunity to attend the Society's meetings when in London, very often being given the greater honour of election as Fellows, to visit its leading lights (notably Boyle or, in the early eighteenth century, Newton), just as when at home they were eager to send their work to the Society in published or unpublished form, and if the latter hoping for publication in England. In the later seventeenth century, and even more in the early eighteenth, travel between the Continent and England flowed much into England where once it had flowed mostly the other way, and all this increased foreign awareness of the aims of the Royal Society whose reputation continued to be that of a body dedicated to experiment and empiricism in an almost Baconian fashion.

It is worth looking separately at the reactions of natural philosophers in different countries, for the English approach was naturally viewed differently in different cultures. And it is reasonable to begin with the Italians, for the English natural philosophers who founded the Royal Society had earlier looked to the example of Italy (as well, of course, of France) for encouragement and method. In the same way, once the Royal Society had come into being, and especially after the demise of the Accademia del Cimento in 1667, Italian natural philosophers looked to it for inspiration, example, and even direct help, to a certain extent perhaps seeing the Royal Society as the heir of their own academy, with the advantage of independence from princely patronage, and soon many Italian academies began direct and deliberate emulation of the English model. Few Italians visited England in the 1660s and 1670s, a situation to change notably later, and fewer even in this period were received into the Fellowship: only three, namely Count Ubaldini, a visitor to London and Oxford, who, it was hoped, might translate some of Boyle's works into Italian; Malpighi, for the excellence of his writings and his willingness to send them to the Royal Society for publication; and Travagino, who asked for the honour.

But even in these early decades the number of correspondents was great, some twenty during Oldenburg's Secretaryship. Of these some few had visited London, notably Magalotti, Secretary of the Accademia del Cimento who had much to do with compiling its *Saggi* (1667); he remained in correspondence for a number of years, sending much news of the Florentine learned world. The remainder were drawn into correspondence because they sympathised with the aims of the Society, as far as they understood them, or because they felt intellectually isolated. Some, like Travagino, approached the Society by sending their books, while others, like Malpighi, were approached by Oldenburg because news of their work had reached the Society, often through other Italian correspondents. They came from many parts of Italy, but particularly from Bologna, Venice and Naples, all flourishing intellectual centres. Relations between Bologna and the Royal Society were particularly close after 1667, a year in which many correspondences began.[7] It cannot be chance that the only work by Boyle published at this time in Italy, his *New Experiments to make Fire and Flame Stable and Ponderable*, was printed at Bologna in 1675, only two years after its publication in England. The Bolognese natural philosophers were on the whole inclined towards the empirical approach to nature characterised by the Accademia del Cimento and by the Baconianism which they thought to be the prevailing doctrine of the Royal Society. Naturally, once the Society began its sponsorship of Malpighi's works, he and his associates were especially favourably disposed towards it. Indeed, it was in large part the Society's direct encouragement which kept Malpighi assiduously at work on experimental biology. And it was largely through him that his friends and associates sought approval for their work, mostly as it happened in experimental physics, seeing the Society as at least potentially a magnificent and beneficent patron of experimental learning. Through Malpighi, Oldenburg heard news of the work of other Italian workers, including something of the astronomical work of G. D. Cassini, who was, however, to migrate to Paris in 1669 (after which his place was in the French world). Others sent their own books for comment and approval either directly or through Malpighi, the former soon becoming in their own right correspondents of Oldenburg. Of these, an example is Montanari, a great Anglophile, who not only sought news of chemical experiments made in England and particularly news of Boyle, but went so far as to procure an English dictionary and grammar in order to read English books, as few Continentals did.[8] Like many other correspondents from many countries, Montanari hoped that the Royal Society would pass judgement upon the probable truth or falsehood of his conclusions, but in this he, like others, was to be told that the Royal Society refused on principle to adjudicate in such matters.[9]

Outside Bologna there was no such network of interest, but many individuals initiated correspondence by writing directly to the Society, usually sending examples of their own works. So Travagino, already mentioned, a minor astronomer and would-be alchemist (he thought he had once made silver, but honestly admitted that he could never repeat the process), as well as organiser of a short-lived academy in Venice, initiated a long correspondence by sending (1667) a copy of his *Idea seu synopsis novae et physicae*, being at pains to stress the importance which in this work he placed upon experiment, for which reason he hoped for the Society's approval.[10] His letters were usually read at meetings of the Society, but, although his request for election was granted, the Fellows remained doubtful about both his achievements and his genuine commitment to true empiricism, as they saw it. The Fellows often more whole-heartedly approved of others never elected: of Borelli, a former member of the Accademia del Cimento, with whom there was some correspondence between 1669 and 1672 and whose works were highly thought of, or of Tommaso Cornelio of Naples, from whom information was sought about the tarantula, and who assured the Society that the spider was not the cause of the dance associated with its name.[11] In Rome there was at this period Francesco Nazari, founder of the *Giornale de'Letterati*, who drew on the *Philosophical Transactions* and on direct correspondence with Oldenburg for news of the English learned world, usually seeking items of an empirical nature.

This Anglo–Italian association and its inclination towards empiricism was to be maintained in the later seventeenth and early eighteenth centuries, with more Italians being elected to the Fellowship and more Italians travelling to England. Among those elected in this period – fourteen between 1677 and 1703, twenty-seven during Newton's Presidency – were not only half a dozen visitors and diplomats but a new generation of natural philosophers, like Guido Grandi, Marsigli, Guglielmini, Baglivi and Bianchini, while ironically, Viviani, with whom Oldenburg had failed to make contact in 1661, was elected in 1696. Marsigli of Bologna, a pupil of Malpighi and of Montanari, was particularly influenced by the idea of the Royal Society's empiricism.[12] He was elected F.R.S. in 1692, having, the previous year, written to the Society about his plan for 'an anatomy' of the Danube, while in 1700 he dedicated the *Prodromus* of his important *Opus Danubialis* to the Fellows. He was to regard his Istituto delle Scienze of Bologna, which he founded in 1714, as modelled on the Royal Society and based firmly on an empirical approach to nature; when it failed immediately to conform to his ideals he left Bologna for France and its Académie Royale des Sciences. But in fact he despaired too soon, for the Istituto revived and flourished. (In 1728 the academicians were to repeat Newton's optical experiments under the lead of F. M. Zanotti.) In the early years of the eighteenth century, many Italian natural philosophers were or would like to have been Newtonians, had it not been for the religious restrictions which weighed heavily on universities and academies. Just as, a century earlier, men had lectured on Ptolemaic astronomy while privately espousing Copernicanism, so now Guido Grandi lectured on Cartesian natural philosophy while privately preferring the Newtonian; Nicola Cirillo, Professor at Naples, described Newton's prism experiments while treating Newton's conclusions from them as mere hypotheses; Poleni was said by his contemporaries to be familiar with Newton's ideas, although he never mentioned them in print; Jacopo

Riccati, a mathematical writer and contributor to both the *Giornale de'Letterati* and the *Acta Eruditorum*, published nothing in his lifetime on Newtonianism, but left at his death an incomplete treatise on physics and cosmology (published posthumously in his collected works) which is filled with experimental philosophy, much of it Newtonian; and Galiani, to whom Newton had destined a copy of the second edition of the *Principia*, wrote privately in praise of Newton to William Burnet (F.R.S. 1705) but published nothing Newtonian. Only after Newton's death did it become easier for Italians to praise Newton openly.

But it must not be thought that all Italians accepted Newton's work uncritically. Giovanni Rizzetti insisted that, since he could not repeat a number of Newton's experiments, Newton's theory of colours must be false.[13] Rizzetti not only wrote about this to a fellow Italian, a letter which came into Newton's hands, but also to the Royal Society directly. The result was that at Newton's instigation Desaguliers repeated the offending experiments with some variations in 1722 and later published an account of them in the *Philosophical Transactions* after Rizzetti repeated his claims in 1727, for Rizzetti remained unconvinced. But his criticism, be it noted, was not of Newton's use of experiment but of Newton's experimental procedure, a very different thing.

About Dutch scientists there is little need to speak, for the names of most of those who were in touch with the Society have occurred frequently in the preceding pages. Huygens had been in close touch with the Royal Society since very early days, first through Sir Robert Moray, then through Oldenburg. He was always in favour of the Royal Society's empiricism except when it moved from the practice of experimental physics to the justification and 'proof' of theory by experiment. He was of course to spend much of his active life in Paris, and moreover was to be cut off from the Society first by Hooke's quarrel over the application of a spring balance to watches, then by Oldenburg's death. He had nothing to say beyond a polite Latin acknowledgement to Grew's letter inviting correspondence.[14] He never accepted Newton's theory of light and colours as proven and, as is well known, although he was much impressed by Newton's *Principia* and admired Newton as a natural philosopher, he declined once again to accept Newton's theories. Leeuwenhoek, on the other hand, as noted above (Chapter 7), continued after Oldenburg's death to regard the Royal Society as his patron, anxious to have his works read, understood and published by the Society. He was to be elected F.R.S. in 1680, the first Dutchman after Huygens (elected 1663) to be so, to be followed by six others between 1680 and 1700, and eight between 1705 and 1726. Many more Dutchmen were correspondents than were ever elected, including De Graaf (d.1672), Swammerdam (d.1680), Constantijn Huygens, and others, so that the Society, especially in Oldenburg's time, was well supplied with scientific news from Holland (and of course from the future Belgium), especially medical news. In spite of the many Dutchmen at Court after 1688 (or perhaps because of it), contacts with the Low Countries were somewhat less vigorous in this period, but the situation was to change in the later years of Newton's Presidency, especially after 'sGravesande's visit in 1715, for as noted above (Chapter 8) he whole-heartedly adopted the empirical outlook of the Royal Society as expressed by Newton, especially in the physical sciences, and in this he was to be followed later by Musschenbroek. The works of these two authors spread the empiricism of the English natural philosophers to all Europe in the middle of the eighteenth century.

Since Oldenburg was German, there was inevitably more correspondence with Germany during his Secretaryship than later, not so much because he sought it out as because Germans, especially perhaps German medical men, expected him to be sympathetic to their attempts to get in touch with the Royal Society. The reputation of the early Royal Society was very high, although not always totally well informed: it was mainly Germans to whom Oldenburg had to write rebukingly to say that the Society was concerned exclusively with *natural* philosophy, that it avoided mysticism, and that it did not profess to judge any disputes about the interpretation of experiments. Thus when Dr Hanneman of Buxtehude 'desired ... judgement on the matter of sanguification' in 1672 the answer was[15]

> that it is not their [the Society's] custom to be hasty in delivering their judgement in any philosophical matters; but that all things of that nature are committed by them to observations and experiments frequently and carefully made.

Oldenburg found no difficulty in replying promptly to the request of Sebastian Wirdig, Professor of Medicine in Rostock, who in 1673 sought the Society's judgement on his system of medicine, knowing well what the collective view of the Society's members would be. Physicians and astronomers (notably Hevelius – but Oldenburg initiated this correspondence)[16] were the most frequent correspondents from German-speaking lands, sending accounts of celestial phenomena and of medical or chemical observations. (The case of Ludolf, the Viennese Imperial librarian and distinguished oriental scholar, is different, for he had corresponded personally with Oldenburg as early as 1659 and may have intended his later letters to be as much for Oldenburg himself as for the Royal Society.) The German learned world on the whole showed the highest esteem for the early Royal Society, even if it did not always completely understand its principles. A notable example of one who did understand them is Nicolaus Witte of Riga, one of whose letters (to Abraham Hill, on preserving wines from freezing) was read to a meeting of the Society in 1669. Witte admired English natural philosophy, which he was able to read in English, above that of the French or Italian, and he eagerly sought English books.[17] Many Latin excerpts from the *Philosophical Transactions* were published in *Miscellanea Curiosa*, the journal of the Collegium Naturae Curiosorum, founded in 1652.[18] Closer to the aims of the Royal Society, indeed partly modelled upon it, was J. C. Sturm's Collegium Curiosorum sive Experimentale, which was established in 1672 with the express purpose of performing 'wonderful' experiments with the usual physical and astronomical instruments of the day. Sturm's account of the work of the first years of his Collegium (published in 1676)[19] was to receive a favourable review in Oldenburg's *Philosophical Transactions*, and, as will be seen, the Society kept in touch with Sturm, although he was never F.R.S.

German medical men, reading abstracts from the *Philosophical Transactions* in their own journal, often wrote to the Secretary with comments, theories and discoveries, hoping for recognition at least for their empirical work. German astronomers were even more assiduous. Of these the best known is Hevelius, brought into the Society's orbit when, on 11 February 1662/3, the Secretary was ordered to write to him in his own name;[20] no sooner had the reply been received (it took a year for this first exchange between London and Danzig) than Hevelius was (1664) elected F.R.S., the first German to be so. After that, and for the rest of his life (d.1687), Hevelius sent astronomical observations and printed books to the Royal Society, secure in its appreciation of his observational work and delightful to have it publicised by the Society. (The only difficulty arose after the publication of Hooke's *Animadversions on the Machina Coelestis of Mr Hevelius* (a Cutlerian Lecture) in 1674, in which Hooke attacked Hevelius for failing to adopt telescopic sights. Hooke was correct in stating that these were more accurate, but, as Wallis pointed out, Hevelius did better with open sights than most astronomers could do with telescopic ones, and the Society deplored Hooke's aggressiveness and lack of tact.)[21] Very many other German astronomers habitually sent their observations of celestial phenomena for comparison with those made elsewhere, asking for the observations of others, English, French or Italian, which they could not otherwise secure, for the Society was widely expected to be the receiver of such information, available in the seventeenth century from no other source. (Hence its frequent publication in the *Philosophical Transactions* for many decades.)

An interesting tribute to German awareness of the Society's empirically orientated point of view came from the young Leibniz. In 1671 he published his *Hypothesis physica nova* in two parts; the first, dealing with 'concrete' or empirically observed motion, he dedicated to the Royal Society, while the second, containing his theory of 'abstract' or theoretically conceived motion, he dedicated to the Académie Royale des Sciences. The view of Pardies, quoted below, suggests that the Society also found it both unoriginal and untrustworthy; indeed Hooke said[22] that he thought that Leibniz 'had not hit it right', while Wallis said that he naturally assented to some parts of the work ('since my own views are the same'),[23] but reserved judgement on the rest, especially on the second part. Oldenburg was forced to be suitably vague in thanking and encouraging the unknown young man, who nevertheless survived this somewhat lukewarn reception which he may have taken to heart, for later communications were either more empirical or, after he settled in Paris and came under the influence of Huygens, mathematical, and he was elected F.R.S. in 1673. He continued to correspond with Oldenburg mainly about mathematics, but when his thoughts turned to more empirical subjects, like his arithmetical machine or his horological invention, he sent accounts to the Royal Society which indeed interested the members, especially when, later, he was able to demonstrate an incomplete model of the calculating machine to a meeting. It is a tribute to his respect for the Society that, returning sadly to a job in Germany, he travelled through London from Paris; this was the time when Collins showed him some of Newton's mathematical papers, a fact destined to generate much acrimony many years later.[24]

It is further worth noting that, in the years between 1676 and 1677 when German chemists were preoccupied with the preparation of various kinds of phosphorus (a word used for any self-shining substance, some requiring exposure to sunlight, some not, the latest discovery being of the modern element), more than one of these sent letters to Oldenburg about their wonders, informing him of chemical progress, of disputes as to discovery and so on, while in the autumn of 1677, after Oldenburg's death, a sample of the new phosphorus was brought to England by J. D. Kraft. When it was shown to Boyle he guessed at the secret of its manufacture and prepared it himself, publishing extensively on it, its method of preparation and its properties (see above, p. 70). The German chemists expected the Royal Society to appreciate such an empirical discovery, as indeed its members did, but, although they spoke and wrote much about the phenomena it displayed, they kept its method of preparation a secret, as Boyle, following Royal Society precepts, did not.

Contact with Germany continued in the later seventeenth century, some fifteen Germans being elected F.R.S. between 1678 and 1700, many of whom were now either physicians or chemists. An example of one with whom Oldenburg had corresponded in 1677 was Franck von Frankenau, a physician and writer on magnetism, who kept in touch with the Society thereafter and was elected a Fellow (nominated by Slare) in 1693. Curiously, the Society did not elect as Fellow a far more likely candidate, one of the few German natural philosophers with whom it proposed to collaborate, namely J. C. Sturm, mentioned above in connection with his experimental college. He was clearly an admirer of the Society, but presumably neither personally acquainted with any Fellow nor sufficiently desirous of the honour to ask for it. Yet when (20 November 1679) the importance of meteorological observations was discussed at a meeting,

> It was desired, that Mr Hooke should write to professor Sturmius at Altdorf requesting him to keep an account there of the variation of the barometer.

Although Sturm in fact did not apparently undertake this task, he was highly pleased to be in correspondence with the Society, and for a number of years he sent many accounts of magnetic and of chemical experiments, of a plan for a new observatory at Nuremberg and, in 1682, a printed sheet asking the 'intelligent and Curious' to send him observations of the variation of the magnetic needle.[25] Contact between the Royal Society and Germany was later spasmodic; during Newton's Presidency relations were naturally somewhat clouded by his dispute with Leibniz, but nevertheless (as noted in Chapter 8) Leibniz's Swiss disciples the Bernoullis were to be elected Fellows. And, be it noted, the very favourable review of *Opticks* printed in the *Acta Eruditorum* was approved of by Leibniz who, like the reviewer, highly praised the book's experimental content.

Relations between England and France varied greatly over the years, close in the very early years and during Oldenburg's period of correspondence, somewhat less later. Six Frenchmen were elected Fellows before 1677, eight between 1681 and 1702, and fifteen during Newton's Presidency. The French always regarded the English Society as wholly committed to experiment, for good or ill. When the Danish scientist Olaus Borrichius visited Paris in 1664–5 – he had passed through London in 1663 and met Boyle and other Fellows of the Society but had not attended any meetings – he was told by Oldenburg's French correspondents of the activities of the Society entirely in terms of experiments performed, noting in his diary with approval a 'Catalogue' (list) of such experiments which had been shown to him and which he carefully copied out.[26] For all that the French natural philosophers, insofar as they projected a corporate image (and this they did imperfectly, even after the foundation of the Académie Royale des Sciences in 1666) were inclined towards Cartesianism, it was a Cartesianism with room for much empirical interest. Ironically, the two most important Academicians in Oldenburg's time were foreigners, namely Huygens, named in the earliest group, already a pensioner of Louis XIV and influential in framing the structure, and Cassini. Both were to be elected Fellows of the Royal Society, Huygens in 1663 (before he was committed to France), and Cassini in 1672, along with only four other Frenchmen before 1680, none of them members of the Académie. All the early Academicians, including Huygens, were 'Baconian' to the extent that they believed in the merit of cooperative experimental labour, as can be seen from the somewhat imperfect registers of the period, but neither Huygens nor Cassini in fact followed this aim to its logical conclusion as others did by publishing anonymously. The Académie had no official journal, although after its foundation in 1665 the *Journal des Sçavans* sometimes published papers of scientific invention, especially by Huygens, making French natural philosophers especially favourably disposed towards the *Philosophical Transactions*. Astronomers were ever-anxious to have their observations published there, witness particularly Auzout in the early 1660s, Boulliaud then and later, Cassini after his move to Paris. The editors of the *Journal des Sçavans* often praised the

Philosophical Transactions and quoted freely from it, its first editor being prepared to state that in England 'la Belle philosophie ... fleurit plus qu'aucun autre lieu du monde', and it was accounts of instruments, experiments and observation which were reprinted in French translation. Oldenburg was in touch with many correspondents in France, especially Paris, and tried to encourage them in empirical interest. In turn the French responded with keen interest, even if not always with acceptance, and usually expressed more or less faith in observations or experiments, if not with conclusions drawn from them. Not many were as open minded as the Jesuit Pardies (whose *Discours du mouvement local* of 1670 was published the same year in England, translated by Oldenburg) when asked in 1671 for a comment on Leibniz's theory of motion, who drily remarked,[27]

> vous me dispenserez s'il vous plaist de vous en dire mon sentiment. Je ne ferois pas la mesme difficulté a l'egard du livre de M. Wallis ou de M. Wren si je les avois lus, car je suis deja fort prevenu de l'excellence de leurs esprit et deleur profonde erudition.

This is a strong endorsement; Pardies was more sympathetic than most to English aims and ideals, but most French natural philosophers did rate their English colleagues very highly, even when they did not agree with their conclusions. Pardies was singular in being so open minded as to have his early scepticism overcome by the force of Newton's experiments and his patient explanations of his optical 'doctrine', becoming a convert to Newton's doctrine or theory of dispersion as early as 1673. Neither Huygens nor his fellow Academicians were ever to do so, although most esteemed his experiments.

Relations between England and France were less close after Oldenburg's death, lacking the personal connection until Sloane became active in the 1690s, but nevertheless four Frenchmen were elected to the Royal Society in the 1680s (admittedly two, Justel and Hautefeuille, being then in England), while four more were elected between 1695 and 1702, and contact did continue. Cassini continued to send his observations in the years after Oldenburg's death, some of which were published by Hooke in his *Philosophical Collections* and then by successive editors in the *Philosophical Transactions*. Cassini could, of course, merely have sent his observations direct to Flamsteed with whom he had been in correspondence since Flamsteed's appointment as Astronomer Royal in 1675, but he evidently valued the approval of the Society as a whole as well as the greater publicity which he achieved by publication in an

English journal. The astronomer Olaus Roemer, Danish in origin but now domiciled in France and a member of the Académie Royale des Sciences, visited London in the spring of 1679, calling on Hooke and on Flamsteed, with whom he later corresponded; he was evidently greatly pleased by this personal contact with English astronomers whose work he admired.[28] Less favourably disposed, although equally empirically minded, were the views of Mariotte, physicist and prominent member of the Académie Royale des Sciences, himself a dedicated experimenter. He followed many of the precepts of English empiricism, as his many experimental works and his discussion of method in his *Essai de logique* (1678) clearly reveal. Since, as has been recently discovered,[29] Mariotte read English and was probably the translator of some issues of the *Philosophical Transactions* which were read to meetings of the Académie; there is every reason to suppose that he was familiar with the work of the members of the Royal Society (including that of Boyle, although he does not mention his name in his own work on the compressibility of air in 1679). He seems to have had no great admiration for English experimental technique, and certainly he criticised severely the early optical work of 'le savant M. Newton' on experimental grounds, finding, as he claimed, major errors in Newton's results.[30] Yet in 1683, when Aston as Secretary wrote directly to him asking for news, Mariotte expressed pleasure at the invitation, apparently esteeming the Society as a whole even while criticising one of its Fellows (not yet a particularly eminent one, it must be recalled). Long after Mariotte's death the reformed Académie Royale des Sciences was to discuss the matter again, while in 1713 Leibniz asked in the *Acta Eruditorum* whether it was not time to settle this dispute. Newton then took up the challenge through Desaguliers, who carefully repeated Newton's experiments with some variations, reporting his success at Royal Society meetings in 1714 and publishing the results in the *Philosophical Transactions* for 1715. By then the atmosphere of the Académie was once more favourable to the English on the whole: the English reciprocated by electing fifteen French Fellows between 1706 and 1726.

The most anglophile academician in this period was certainly E. F. Geoffroy; his visit to England in 1698 (Chapter 8), his consequent election as F.R.S., which he called 'an honour I esteeme above all things in the world',[31] his attendance at meetings, and his command of English all predisposed him in the Society's favour. After this visit he corresponded frequently with Sloane, sending news of experimental work performed at the Académie Royale des Sciences and welcoming news of what occurred at the Royal Society's meetings. He was perhaps all the more eager since in the 1690s experimental science within the Académie was somewhat overshadowed by mathematics and the mathematical sciences, a situation which was to change with the reorganisation of the Académie in 1699, when experimental chemistry, of which Geoffroy was an able practitioner, was to have an important place. It is tempting to assume, as many historians have, that Geoffroy had at least something to do with the double nomination in 1699 of two English natural philosophers to the Académie Royale des Sciences: Sloane as a *membre correspondant* (the fact that it was with Jacques Cassini, an astronomer like his father, is not necessarily relevant) and Newton as one of the first of a new class of *associés étrangers*. Certainly it was Geoffroy who wrote to inform Sloane of the honour awarded to him and to Newton, promising, because of 'the gratitude that I conserve for the honour they have confer'd to me', to send the Royal Society news of 'all curious things that will be found here', a promise he duly fulfilled.[32] There can thus be no doubt that Tournefort's earlier friendship with Sloane was ultimately the cause of closer connections between the Royal Society and the Académie Royale des Sciences in the 1690s and early years of the eighteenth century.

Although foreign opinions were generally favourable to the Royal Society's experimental way, as noted above some Continental natural philosophers had, at best, mixed feelings. True, Mariotte criticised particular experiments devised and described by Newton, not his use of experiment in natural philosophy. Attacks on Boyle in the 1660s had been at least partly concerned with his interpretation of experiment, not wholly with his devotion to it. But it must not be forgotten that, when Boyle died, both Leibniz and Huygens in mutual correspondence while deploring his loss, declared that he had at least to some extent wasted his talents in 'only' performing experiments and in neglecting to develop a comprehensive theory or to consider deeply whether his experiments alone were the reason for the adoption of his conclusions.[33] Yet Huygens and Leibniz were both proud to be Fellows of the Royal Society. Many Cartesians, English and Continental, accepted the Society's empiricism with the reservation that, admirable though the experimental philosophy was, true systematic and theoretical natural philosophy demanded the use of the Cartesian rational or mathematical way, based utlimately on *a priori*, rational metaphysical principles. Huygens and Leibniz were by no means singular in this.

Certainly, by no means all the English Fellows were as totally committed to purely experimental reasoning as Boyle or the younger Newton, however committed to belief in the importance of seeing or hearing accounts of the performance of experiment. Many sympathised, in whole or in part, with a more Cartesian mode of reasoning, which experiment assisted but did not dominate. Many, however deeply and truly empirical in belief, found an acceptance of experiment as the crucial test for theory impossible, clinging tenaciously to their own hypotheses (however much they recognised and rejected the hypotheses of others). Thus Hooke in his 1670s controversy with Newton over the nature of light and colours argued that Newton's experiments could be made consonant with his (Hooke's) theories which were accompanied by hundreds of experiments, as he claimed. Many lesser Fellows were outspokenly in favour of the precedence of theory over experiment. Late in 1677 a curious interlude was provided to the Society's normal proceedings by Oliver Hill, a theological writer proposed for election only on 8 December 1677 (on Boyle's recommendation), although he was present at the meeting of 6 December and possibly even at that of 15 November. Shortly after his election (13 December) he spoke fervently in favour of theoretical rather than empirical reasoning. As the minutes record,

> he thought, that it was going very much about to begin with experiments and end with theory, and affirmed, that he had ... from a theory, of which he was master, of the nature of the air and of mercury, and from principles of his own invention, plainly showed the reason of all the experiments, which had been exhibited and discoursed at the [previous] meeting, and why things happened so and not otherwise,

offering to bring in his theory to the next meeting, which

> the President [Sir Joseph Williamson] desired him to do, although he was acquainted, that the method and business of the Society were very different from those he propounded; it being their aim rather to be directed by the operations of nature duly observed, than by theories not built upon a sufficient and unquestionable foundation of observations and experiments.

This criticism is perhaps unfairly severe (though conventional), for his intervention had arisen from comments arising out of the minutes of the previous meeting, which had provoked 'a debate ... concerning the cause of the phaenomenon of the barometer', during which several fairly speculative theories had been advanced by Hooke, Jonas Moore and

Grew, although there had also been proposals for experiments to test these theories from several members, all of whom alleged the existence of confirmatory experiments for their particular theories. Hill was by no means discouraged: he maintained

> that we ought to be ruled by a theory in the making all our experiments, yet he would be understood to have the theory founded upon previous experiments,

a somewhat muddled conclusion, nor is it clear whether his theory, of which he read an account on 20 December, involved experiments before or after the fact. Hooke, according to his *Diary*, would have nothing to do with it, and he derided King, whom he privately denominated 'an asse' for admiring Hill's statement – not surprisingly in view of Hill's argument that air had no gravity, a 'fact' he claimed to have proved by experiment, an experiment which failed before the Society, although he later (17 January 1677/8) reported that it had succeeded in private. He was not given a further chance to discuss the matter, although he was subsequently asked to report upon a discourse on Hermetic philosophy (by Edward Smith of Chichester) which he duly did. After this his name vanished from the minutes, although he continued to be a Fellow.

Hill was clearly a man of little intellectual weight, but by no means alone in taking the unpopular side in the long drawn out discussions deriving from Boyle's insistence that he had *proved experimentally* both the weight and the spring of the air. Books by George Sinclair (1669) and Matthew Hale (1674) had both been taken seriously, even though Boyle had not thought it worthwhile to combat them at length, and the Society ignored them except for reviews in the *Philosophical Transactions*. But these were attacks only upon some results of the experimental way, not attacks on the method itself. Yet it should not be lightly assumed that all the world did agree that the experimental was the most desirable method in natural philosophy. Many, even those who were prepared to praise the achievements of Boyle and of Newton, regarded experiment as but a minor part of rightly reasoning in natural philosophy, pre-ferring to start with rigidly logical principles.

The serious objectors to the experimental way ranged from those who, while supporting experiment in conjunction with theory, preferred the Cartesian, rational or mathematical way of constructing theory, to those who, failing to repeat experiments described in print by Fellows such as Boyle and Newton, rejected – and not without reason – the conclusions based upon these experiments, to those who flatly and totally rejected the precedence the Society and its leading members claimed to give to experiment when devising hypotheses and theory. Two philosophers not unversed in practical or empirical matters who both found themselves unable to accept experiment as of fundamental importance in the framing of theory and hypothesis were Spinoza and Hobbes. When Spinoza (and in this he agreed with Descartes) constructed his philosophical theories his criterion was rigour, and he believed that his ethics was indeed constructed *more geometrico*, that is as rigorously as the propositions in Euclidean geometry. Moreover, again with Descartes, he firmly believed a scientific proposition to be sufficiently demonstrated when it can be shown to be a strictly logical consequence of a set of intuitively rigorous axioms. So when in 1662 he carefully read Robert Boyle's *Certain Physiological Essays*, in which Boyle supported the view that a scientific theory should be demonstrated by showing that it holds empirically, Spinoza was stimulated to a long and careful refutation of such a view in a series of letters to his acquaintance (and Boyle's literary assistant) Henry Oldenburg, who replied at length, as instructed by Boyle.[34] To Spinoza, common sense observation of phenomena was sufficient empirical evidence while elaborate experiment was no real guide, especially when dealing with such a complex and empirically remote problem as the particulate structure of matter. Indeed, Spinoza could not bring himself to understand the nature of a chemical reaction nor the difference between a chemical compound and a physical mixture, problems soluble, as Boyle well knew, from empirical evidence obtained by careful experiment, which Boyle therefore valued above all. Spinoza, rejecting such procedure, was left perplexed. Theory based firmly on experiment, dear as much to Boyle as to Newton, was in fundamental opposition to reasoning in the way favoured by Spinoza. Therefore the long exchange of letters between Oldenburg, acting for Boyle, and Spinoza petered out in fruitless lack of understanding.

Hobbes also attacked and refused to accept Boyle's empirically based natural philosophy, questioning both the value and the facts of his pneumatic experiments; in doing so he rejected by implication the whole attitude of the Royal Society, just forming as he wrote his attacks. The criticism of Hobbes is more closely tied to Boyle's actual experimental work than that of Spinoza, chiefly because Hobbes had thought deeply about such problems as the structure of matter and physical forces.[35] So when he attacked Boyle's first pneumatic experiments, he denied the validity of Boyle's conclusions (such as that an animal in the evacuated receiver died from absence of air) because these conclusions contradicted the principles which Hobbes had long ago laid down in his *De Corpore* (1655). Similarly, Hobbes denied that Boyle had proved that air possessed elasticity (or spring) because this contradicted his own principles of motion and 'restitution'. Hobbes insisted that his principles could explain all Boyle's conclusions adequately, and moreover censored Boyle for limiting his conclusions to physical nature, ignoring politics and ethics (in the way that the Royal Society had always done and was always to do). On stronger ground, Hobbes criticised the less than perfect working of the airpump, which, especially the first model, was inclined to leak perceptibly. This criticism was fair enough as far as the technicalities went, but to Hobbes the imperfection in the equipment totally invalidated both Boyle's experiments and his conclusions. For, while to Boyle it was an easy matter to make allowance for the imperfections, which he amply recognised, or to correct them (as by designing a manometer to show how much air had been removed in the first place and how much leaked in during the course of the experiment), and in any case did not expect either perfect consistency or perfect results when examining nature, to Hobbes, for whom no allowance could be made for human imperfection when Nature was being investigated, all this merely confirmed the superiority of the rational as against the empirical approach. Boyle was entirely conscious of the problems inevitably associated with the experimental method, as he had revealed in his essay of 'Unsucceeding Experiments' in *Certain Physiological Essays* (1661), and he clearly found it difficult to cope with Hobbes's demands for experimental perfection. In 1664 Hobbes was still analysing others' experiments which seemed to demonstrate air's compressibility and refuting the conclusions by means of his own principles of body and motion; he further clouded the issue by relating these problems to the mathematical principles whose complexities had embroiled him with John Wallis in the earlier 1660s.[36]

Now the criticism of both Spinoza and Hobbes was directed explicitly to the work of Boyle (and of Wallis, in the case of Hobbes). It thus might appear to be irrelevant to the Royal Society as a whole; after all the criticism of Newton's optical experiments in the 1670s by Huygens, Linus and others, or indeed the criticism by Linus of Boyle's pneumatic arguments in 1661 was personal and not directed against the Society as a whole. But Spinoza and Hobbes both criticised not only Boyle's experiments and his experimentally based conclusions but his very method of using experiment to establish theoretical conclusions, rather than using experiment to confirm and perhaps extend principles derived from rigorous logical thought. And since Boyle was a leading member of the early Royal Society, and in these early days the work of individual members was closely identified with the Royal Society, and fairly so when Boyle's experimental way was pre-eminently that proclaimed as the Society's way, to criticise Boyle's method and results was to attack the method and purpose of the Royal Society itself.

The absence of general principles other than those of the experimental way was seen by many contemporaries as a profound weakness in the fundamental concepts and working of the Royal Society. On the whole, seventeenth-century natural philosophers still valued systems and found experiment without such principles seriously incomplete. Even many of the Fellows agreed, finding a framework necessary for their experiments and, tacitly or implicitly, accepting something more than empirical investigation alone as necessary to advance natural philosophy. Historians, like contemporaries, have often taken the will for the deed and assumed that such men as Boyle, Wallis, Hooke, Petty, Newton and Halley all followed a mainly experimental and empirical method, because they claimed to do so. But the extent to which they in fact did so must be judged to vary from one to another and even from one situation to another. Boyle, certainly, intended to ground all his investigation into nature on experiment and believed that he had succeeded in this endeavour. But his corpuscular philosophy, of which he was so proud, although convincing, did not, as it seemed to many, require so much experimental evidence as he laboriously produced in its support, and with hindsight many historians have judged that he had first come to believe in its truth on purely rational grounds, before he had available the great body of evidence which he adduced in its favour. It is necessary, however, to realise that to him, as to Newton, experimental evidence was of truly great importance, and the way of looking at nature practiced by these and other English natural philosophers did place far more emphasis on experiment compared with *a priori* theory than was the case with more rationally minded natural philosophers, especially many Continental contemporaries. These were apt to declare emphatically that the weakness of Boyle's empiricism was the cause of what they regarded as an almost catastrophic failure, a failure to create any system of chemical philosophy – for it was Boyle's chemistry which had the most lasting impact on his later contemporaries rather than his corpuscular philosophy, by then taken for granted as a variant of what was the prevailing 'mechanical philosophy' of the later seventeenth century. Boyle was aware of his lack of systematic chemical theory, but did not regret it, for he thought that his achievements – the making of chemistry a genuine part of natural philosophy, its divorce from mysticism on the one hand and from entirely practical medical utility on the other – were sufficient. His way of looking at chemistry in terms of simple substances and the determination of their identity by means of empirical chemical tests, systematically and almost uniquely practised by him, permitted a useful and reproducible view of chemical composition which was to be adopted by many who refused to accept his empirical philosophy. In the final judgement of history, his empiricism must be taken to have led to more lasting conclusions than the systems and principles of the rationalists; it is no accident that much eighteenth-century French chemistry followed Boyle's lead even while retaining principles of which he would have disapproved, nor that Lavoisier began his road towards the new chemistry by repeating many of Boyle's experiments.[37] And since Boyle's name was so closely linked to that of the Royal Society, the fortunes of his reputation came close to the changing reputation of the Society itself and of its adherence to the experimental way.

Criticism of the empiricism of the Royal Society and of its Fellows by no means always came from philosophers and natural philosophers. The very success of the Society, the fame of its Fellows, and the evident prominence of its publicity caused its ideals to permeate both the learned and the unlearned world, the universities, the Court and genteel society. Sprat had publicised the Society's interest in the language of exposition; in response, the literary world felt entitled to comment upon the Society's aims and activities (which Sprat had also publicised). In an age of satire such as the later seventeenth century the Royal Society was an easy target. The university world and the College of Physicians attacked the Society for trying, as they saw it, to dominate scientific learning and perhaps to take over their academic privileges. They strongly resisted the Royal Society's claim to have discovered the true – that is the experimental – method of proceeding to investigate nature. To satirists and wits the idea that men of learning, many of them gentlemen, should spend their time on the investigation of such lowly things as the very small, whether creatures or parts of matter, was laughable, and this rendered the Society's empiricism in general and its experimentalism in particular wholly derisory. Indeed it was all too easy to describe many of the experiments made at Society meetings as trivial, while its practical interest in new machines and histories of trades was thought unsuited to gentlemen or learned men. From the beginning of its formal existence the Society and its members were laughed at by many for such seemingly purposeless activities. Well known is the story, quite well attested, of their patron King Charles II laughing at the gentlemen of 'his' newly chartered Society who spent their time only in weighing air and devising obviously impractical ships (that is Petty's 'double bottom' boat, a catamaran).[38] It was a strong lead for courtiers to do likewise, unless they were, like among others, Sir Robert Moray and Sir Paul Neile, Fellows of the Royal Society.

The earliest example of this attitude is probably the satire on 'The Choyce Company of Philosophers and Witts who meete on Wednesdays Weekly at Gresham College' by an unknown, but obviously well-informed versifier, so well informed that in modern times the satire has been often lost.[39] Here the pretensions of the Society in claiming to have found a 'new' philosophy to replace the well-established old philosophy and to have done so by empirical means are roundly derided. Derisively the satirist sets out their aims:

> By demonstrative Philosophie
> They Playnly prove all things are bodyes,
> And those that talke of Qualitie
> They count them all to be meer Noddyes.
> Nature in all her works they trace
> And make her as playne as nose in face ...
>
> Therefor in Counsell sate many howres
> About fileing Iron into Dust
> T'experiment the Loadstone's powers;
> If in a Circle of a Board they strew it,
> By what lines to see the Loadstone drew it ...

And so on for many verses, the later ones succinctly describing pneumatic experiments, the histories of trades, experiments on agriculture, Wilkins' *Universal Character*, Evelyn's plea against coal smoke (*Fumifugium*), the search for improvements in navigation, and the repetition of Bacon's experiments on the preservation of natural substances by freezing. The unknown author was familiar with what was done at meetings, perhaps had even attended one or more, but equally clearly had little or no interest in the purpose of the experiments and discussions, and (unlike, say, Pepys) found them uncongenial. Much information and misinformation about what went on at meetings was available in the coffee houses of London, where Fellows often went after meetings, to be joined by non-Fellows in serious discussion and/or gossip. There is no clearer testimony to the knowledge of what went on within the Society and the derision to which it might be subjected by non-members than the declaration of Sir Philip Skippon in a letter to John Ray of December 1667. He reported that[40]

> The Effects of the Transfusion are not seen, the Coffee-Houses having endeavoured to debauch the Fellow [the subject, one Arthur Coga] and so consequently discredit the Royal Society and make the Experiment ridiculous.

Most satirists were to take the trouble to familiarise themselves, at least by report, with details of experiments performed. The two chief points of attack were, as they continued to be, the deriding of elaborate experiments used to prove what was taken to be common knowledge, and the deriding of the practice of gentlemen who went to school to tradesmen or artisans. Most laughable of all to the satirists was the use of elaborate, complex and expensive apparatus – telescopes, microscopes, airpumps – to demonstrate the obvious (that air was necessary for life and fire), the improbable (the nature of the surface of the moon, the existence of 'eels' in vinegar) or the 'obviously' unimportant (the anatomy of the flea, the physical structure of a hair). This followed a satirical tradition as old as the Greeks, which believed that those who investigated the natural world were themselves impractical, and like Thales fell into wells because they were gazing at the stars. Moreover, although most people were more or less credulous as regards astrology and alchemy, scepticism was widespread, and it was not thought that learned men should indulge in their practice. The public at large was sufficiently confused about the differences between astronomy and astrology or alchemy and chemistry to make little distinction between them. So Sidrophel, the credulous astrologer of Samuel Butler's *Hudibras*, who was in fact modelled upon the well-known astrologer and almanac maker William Lilly, first active in the 1640s, was later taken to be modelled upon the astronomers of the Royal Society. Almost all of Butler's work was in the form of verse satire, and as a satirist he naturally concentrated upon the follies of men. In other verse satires[41] (most of which circulated in manuscript) he particularly made fun of the pretensions of natural philosophers and the 'obvious' defects of their observations and experiments. So in 'The Elephant in the Moon' he describes how the learned virtuosi 'saw' such a beast with their telescope, until a servant (illustrating in the best and oldest tradition of satire, the common sense of the common man) pointed out that there was a mouse in the telescope tube. In other verses whose title proclaims them to be a direct attack on the Society, namely 'On the Royal Society', Butler laughed at those who

> ... measure wind and weigh the air
> And turn a Circle to a Square ...

Here the message is clear: the experiments performed by the virtuosi were foolish, foolish conclusions were drawn from them, and all experimental investigations were trivial and useless. Like the learned of all ages, the natural philosophers of the Royal Society were impracticable and lacked common sense; hence their activities must be laughable to all sensible men, and since their most notable activities were experimental, experiment itself must be a laughable activity. There was little the Society could do to counter such satire, and if the Fellows thought that they could in any measure do so by means of Sprat's *History*, they deserved more derision: for Sprat was too intensely serious and solemn a writer to do more than to protest that experiments were a noble means to the attainment of knowledge and necessary to reinforce and correct common sense as well as to acquire new knowledge, while histories of trades were of use to mankind and hence not beneath the dignity of any true natural philosopher. In any case, it is impossible effectively to counter satire, for a serious reply can only provoke more laughter.

The publication of Sprat's *History* in 1667 was in fact largely ignored by satirists. But, combined with Glanvill's similar claims for the new philosophy (p. 52), Sprat's work offered a target for attack to those who disliked the new learning, not because it was trivial but because it tried to displace the old learning. These critics were mostly learned but deeply resistant to new forms of thought; they were not concerned with the validity of the experimental way but with metaphysics and the proper rôle of learning. Above all they were concerned with what they saw as the protection of religion. Sprat had anticipated or was apprised of their points of view and had done his best to answer their criticism before its publication. But in their criticism the detailed experimental work of the Society was largely ignored, and for that reason the attacks by such men as Henry Stubbe have no place here, especially in his case because he conducted direct war with Glanvill and not directly with the Society, although he did, it is true, regard Glanvill as a fair representative of the Society.

As the work of the Royal Society proceeded, so the satirists enlarged upon their satires, attacking microscopy as publicised by Hooke and Leeuwenhoek, medical experiment as exemplified by blood transfusion attempts and inventions like Petty's double bottomed boat. Thus in 1669 the Oxford University orator, Dr South, in a general public attack on the Society, spoke of its members as doing 'nothing but magnificare pulices pediculos et semetipsos'.[42] Microscopy indeed soon replaced telescopic astronomy as the most apparently ludicrous of the new branches of natural philosophy, being even more derided than experiments with the airpump. One of the most famous satires on the Fellows of the Royal Society was by the playwright Thomas Shadwell (1642?–92) who, unlike Butler, must have been familiar with the Society's existence and activities during all his adult life. It is not clear how he did learn in detail of the Society's affairs, except that, as noted above, the coffee houses must have provided excellent sources of information in an age when conversation there was a favourite recreation (they provided places where views could be exchanged and news gathered much as London clubs were to do in the nineteenth century), and the Royal Society was of importance in London life. Shadwell and Dryden had already exchanged satirical attacks, and in attacking virtuosi Shadwell may partly have aimed at his literary enemy; if so, the aim was somewhat wide of the mark, for although Dryden was an Original Fellow, he was never closely involved with the Society, even being expelled for non-payment of dues in 1666. Shadwell's attacks in *The Virtuoso* of 1676 had a wide impact for it was to become an immensely popular play, known and performed for many years. These attacks were aimed not at the working members of the Society, strictly speaking (whatever Hooke may have thought),[43] but at the gentleman-virtuoso who certainly represented one element of the Society's membership. Sir Nicholas Gimcrack, the virtuoso of the title, spends all his time looking at the blue of plums through a microscope. He is a classic figure of fun, one more example of the learned man who is totally impractical, absent minded, over-enthusiastic, more interested in microscopic creatures than in his fellow men or even his wife, who squanders his substance on instruments and inventions which fail to contribute anything substantial to the good of mankind, and in any case fail to work. It is a clever piece of satire as well as successful theatre, and it was clearly seen as being directed at the Royal Society; so successful was it that Shadwell's character became the archetype of the foolish learned man, especially the foolish natural philosopher. The word 'virtuoso' became thereby seriously devalued, although of course the type continued in existence through much of the eighteenth century. Nearly twenty years after the appearance of Shadwell's play, William Wotton, ably and knowledgeably defending modern philosophy in his *Reflections upon Ancient and Modern Learning* (1694),[44] which compared ancient natural philosophy unfavourably with that of the 'moderns' of the sixteenth and seventeenth centuries, could speak regretfully of

> the sly insinuations of the Men of Wit, That no great Things have ever, or are ever like to be performed by the Men of Gresham, and, That every Man whom they call a virtuoso, must needs be a Sir Nicolas Gimcrack; together with the public Ridiculing of all those who spend their Time and Fortunes in seeking after what some call Useless Natural Rarities; who dissect all Animals, little as well as great; who think no part of God's Workmanship below their strictest Examination.

This ridicule, Wotton went so far as to say, had discouraged gentlemen from pursuing the original aims of the Royal Society, and he feared for the future. It is a strong tribute to the effectiveness of Shadwell's satire.[45]

Wotton's work properly belongs to the strictly literary quarrel known as 'the Battle of the Books', a long drawn out dispute begun in the seventeenth and continued into the eighteenth century, as to whether the ancients or the moderns had produced the more important contributions to literature. This old dispute among classical scholars was about to be recognised as sterile, since matters of taste were more deeply involved than matters of fact in judging whether anyone had ever exceeded Demosthenes or Cicero as an orator, whether Shakespeare could be compared with Sophocles, and so on. Of course, to criticise one's contemporaries for taking idle pleasure in novelty, for preferring the perhaps trivial new idea to the established solid old idea, was also an established conservative view. Novelties (innovations) were the main target of many like Meric Casaubon, who had been repelled by the works of Glanvill and Sprat precisely on the grounds of their advocacy of novelties, for Casaubon, like many before him, saw novelties as being an inherent threat to belief and as leading ineluctably to irreligion. Unlike most such conservatives, Casaubon noted the importance of experiment as a key element in the defense of modern as against ancient learning by those who supported the new learning. In 1668, in a work entitled *Of Credulity and Incredulity*, he denounced 'modern' authors, that is, all

> They . . . that would reduce all learning to natural experiments or at least would have all learning . . . regulated by them, and those that profess the trade, whether meer Empiricks or other

as men who inevitably lead their readers away from true religion, weakening faith and hence the bonds of society. Of such men, Glanvill and Sprat were, to Casaubon, important exemplars.[46] The old quarrel whose origins lie in sixteenth century delight in new discovery flared up in late seventeenth-century France, while in England William Temple defended the ancients as the greatest scholars, orators and philosophers, arguing that recent discoveries about the world of nature were made by modern pygmies standing on the shoulders of the ancient giants – not that he despised the new, but that he regarded it as less of an achievement than that which had been made by the (ancient) initiators. Wotton, on the contrary, urged the manifestly greater knowledge about the natural world displayed by the moderns to claim firmly that the seventeenth century was more learned and knowledgeable than the ancient, whether its natural philosophers were greater thinkers or not; what mattered was the result, not the mental calibre of those who achieved it. He did not take up the challenge about the dangers of experiment, although he did give examples of cases where experiment had led to new knowledge. Even Wotton, writing informed history of science late in the century, did not care directly to address the issues raised by such men as Casaubon.

But things were changing in the relations between the world of learning in general and the more restrictive world of natural philosophy, the literary world of the early eighteenth century drawing away from the world of natural philosophers. Fewer and fewer literary men were now Fellows of the Royal Society. Many writers saw the natural philosopher as, if not the enemy of literature and the arts, yet as increasingly remote from them. However, the satirists and wits of this period were still conscious that the Royal Society under Newton's Presidency had its place in the London society of the day; they expected their readers to agree and to recognise its preoccupation with experiment, at which it was still fashionable to laugh. This was quite without reference to politics, even the politics of literary feuds and rivalry. The essayists Addison and Steele were Whigs, Pope was a Tory, Swift was first a Whig, later a Tory; the Royal Society was totally apolitical, although it was always ready to respond to Government requests for assistance. (The Royal Society supplied Visitors, a kind of overseeing Board of Directors, to the Greenwich Observatory after 1710, for example.) Addison and Steele, in the very popular *Spectator* and *Tatler*[47] derided the Fellows as being of little wit, dull and without learning, kept out of the real world, that is the political world, because they were distracted by playing with such toys as scientific instruments, more interested in insects, animal experiments and eels in vinegar than in man, society and God. So too Swift satirised the Royal Society (and all learned men) in Gulliver's *Voyage to Laputa* (1726), guying the traditional absent minded-ness of academics and the apparently impractical inventions of the Fellows, describing a man 'who had been eight years upon a project for extracting Sun-beams out of cucumbers'. Like Shadwell's, Swift's satire was so effective that 'to get sunbeams from cucumbers' was still such a well-known absurdity in the late nineteenth century that W. S. Gilbert could employ it in *Princess Ida*. In 1712 Alexander Pope in his turn derided those

> Who study Shakespear at the Inns of Court,
> Impale a Glow-worm, or Vertu profess,
> Shine in the dignity of F.R.S.

again associating divorce from the world and the study of natural philosophy.[48] It was above all the apparent triviality of the study of the very small, the serious contemplation of insects and nematode worms (vinegar eels), and the study of the seemingly obvious characteristics of nature that struck all these early eighteenth-century satirists. So William King's parody of the *Philosophical Transactions* held up to ridicule the obsession, as he saw it, of natural philosophers with the trivial and their ignorance of men's normal and proper preoccupations. His point of view is well illustrated by his regarding Martin Lister's *Journey to Paris* (1699) as a prime example of this absurdity: Lister, he said, was interested only in what King regarded as commonplace natural history and neglected the great works of architecture, art and those activities of man in society which most appealed to normal travellers. (True, Lister was more interested in gardens, in curiosities and in instruments than in architectural details, and in gardeners and librarians than in society people and the court; yet he was, in fact, keenly interested in the customs, manners and conditions of people both high and low, giving an excellent account of the topography of Paris, but of the salons as well.) King's *Transactioneer* (1700) wickedly and exactly caught the spirit of many of the minor natural history articles to be found in Sloane's journal, accounts which King thought to be full of useless and incomprehensible details of nature which, he was sure, could never be important to know. King produced several issues of a clever parody, *The Useful Transactions*, in 1708 and 1709, the title itself a direct attack upon the *Philosophical Transactions*, which, to King and his audience, manifestly failed to live up to the claim to be full of useful information.

Clearly, all this satire was remote from questions of political association; Whig and Tory alike could laugh at Sloane and even Newton, deriding experiment, observation and the use of scientific instruments in equal measure. Sir Isaac Newton, Master of the Mint and President of the Royal Society, was an influential figure in many circles, but it was not until he was dead that ordinary men saw fit to honour him as England's greatest natural philosopher. Pope's famous lines,

> Nature and Nature's laws lay hid in night
> God said, 'Let Newton be!' and all was light

were written in response to a request for a suitable epitaph on a memorial plaque. They were rejected for this purpose, but half a dozen years later Pope liked them well enough to include them in his *Essay on Man* (1733). Whether the climate of opinion had changed now that Newton was safely in his grave, or whether Pope merely wished to preserve the neat lines is impossible to tell. It may be that the Royal Society was no longer such an integral part of London intellectual and coffee house life as it had been for so long, and so was less worth the trouble of satirising. In any case, its former devotion to experiment, as has been shown above, had weakened. When Desaguliers died, experiment at meetings was already a thing of the past, and formal papers might or might not contain detailed descriptions of experiments performed privately, even when they were experimentally based. Only the award of the Copley Medal kept alive the experimental emphasis of early days, and that was increasingly awarded not for experimental *display* but for experimental *research*, whether published or unpublished, which led to important conclusions, the emphasis being on the results and no longer on the experiments themselves. It was clearly now believed that research in natural philosophy could no longer be either made or demonstrated publicly and that experiment was so little of a novelty that it could not catch either professional or amateur attention. In all the many outside attacks which bedevilled the Royal Society in the nineteenth century, there was never any serious emphasis upon experiment as the characteristic feature of the Society's methodology. By then what mattered was the effectiveness of scientific discovery, and no longer the method by which that was carried out, which was taken for granted.

Abbreviated titles

Alle de Brieven	*Alle de Brieven van Antoni van Leeuwenhoek*
Birch, *Boyle*	Thomas Birch, *The Works of the Honourable Robert Boyle*
Birch, *History*	Thomas Birch, *The History of the Royal Society*
BJHS	*British Journal for the History of Science*
BM MS.	British Library manuscripts
Heilbron, *Physics*	J. L. Heilbron, *Physics at the Royal Society during Newton's Presidency*
Hooke, *Diary*	(1) H. W. Robinson and W. Adams (eds.), *The Diary of Robert Hooke M.D., F.R.S. 1672–1680*
	(2) R. T. Gunther (ed.), *Early Science in Oxford*, vol. x, 'The life and work of Robert Hooke, Part iv, Diary, 1688 to 1693'
Newton, *Correspondence*	H. W. Turnbull, J. F. Scott, A. R. Hall and L. Tilling, *The Correspondence of Isaac Newton*
Notes & Records	*Notes and Records of the Royal Society of London*
Oldenburg, *Correspondence*	A. R. and M. B. Hall (eds.), *The Correspondence of Henry Oldenburg*
Pepys, *Diary*	R. Latham and W. Matthews (eds.), *The Diary of Samuel Pepys*
Phil. Trans.	*Philosophical Transactions*
P.R.O.	Public Records Office
Sprat, *History*	Thomas Sprat, *The History of the Royal Society of London for the Improving of Natural Knowledge*

For bibliograpical details, here and in the notes, see the Bibliography (p. 193).

Notes to the text

CHAPTER 1

1 *The Record of the Royal Society*, 4th edn., pp. 289 and 311–12.
2 Letter to John Herschel, 10 December 1839, Royal Society MS. HS 9,
  no. 198, and *Letter addressed to the Earl of Rosse* (1848), esp. pp. 13–14.
  Cf. M. B. Hall, *All Scientists Now*, p. 232, note 9.
3 Royal Society MS. DM 5.11 (1). For the ascription to William Neile, see
  Hunter and Wood, 'Towards Solomon's house'.
4 Heilbron, *Physics*.
5 M. B. Hall, 'Science in the early Royal Society', and 'Salomon's house
  emergent'.
6 Hunter and Wood, 'Towards Solomon's House', reprinted in M. Hunter,
  Establishing the New Science, which contains more on this point.
7 See especially Shapin, 'The house of experiment in seventeenth century
  England', and Shapin and Schaffer, *Leviathan and the Air Pump*. See also, for
  example, Jan Golinski, 'A noble spectacle', and Rappaport, 'Hooke on
  earthquakes'.
8 The latest appraisal is by Miller, '"Into the valley of darkness"'. But it
  should be remembered that from 1710 onwards the Society often acted for
  the Crown in scientific affairs.
9 See Birch, *History*, 6 February 1683/4. The book was *De herniis* by 'Dr
  Munckhausen' [Munchausen], probably Benjamin von Munchausen
  (F.R.S. 1684).

CHAPTER 2

1 Original minutes, quoted in *The Record of the Royal Society*, 4th edn., p. 7.
2 Letter to Richard Norwood, mathematical teacher and practitioner in the
  Bermudas, 6 March 1663/4; Oldenburg, *Correspondence*, vol. ii, p. 146.
3 For the ultra-Baconian view, see Purver, *The Royal Society: Concept and
  Creation*, criticised in *Notes & Records*, **23** (1968), pp. 129–68 by P. M.

Rattansi, C. Hill, and A. R. and M. B. Hall, under the general title 'The intellectual origins of the Royal Society'. For contemporary criticism, cf. the works of Henry Stubbe, directed primarily at the books by Joseph Glanvill and Thomas Sprat rather than at the actual workings of the Society.

4 By Oldroyd, 'Some writings of Robert Hooke'.

5 Sprat, *History*, pp. 61, 83 and 95. A convenient edition is that edited by Cope and Jones with facsimile text. Cf. also Wood, 'Methodology and apologetic'.

6 For attempts to appoint Curators to lift the burden from Hooke see below and also Hunter and Wood, 'Towards Solomon's house', esp. pp. 51–2, reprinted in M. Hunter, *Establishing the New Science*, pp. 185–244, and *ibid.*, pp. 279ff, 'Science, Technology and Patronage: Robert Hooke and the Cutlerian Lectureship'.

7 Birch, *Boyle*, vol. vi, pp. 288–91. Cf. M. Hunter, 'A "College" for the Royal Society', *Notes & Records*, **38** (1984), 159–86, reprinted in M. Hunter *Establishing the New Science*, pp. 181–4. Hunter there also deals with the plans formulated in 1667–8.

8 Oldenburg, *Correspondence*, vol. iii, p. 45, letter dated 24 February 1665/6.

9 Sprat, *History*, p. 434.

10 For details (many in Birch, *History*) see M. Hunter, *Establishing the New Science*, pp. 156–81. A total of £1285 plus the site was pledged.

11 It is to be found in Royal Society MS. DM 5.11 with a partial transcription by Oldenburg endorsed (without naming the author) 'Proposals concerning Experiments' and is printed by Hunter and Wood, 'Towards Solomon's house' (note 6). For some of Neile's scientific activities, see Oldenburg, *Correspondence*, vols. iii–vi.

12 18 June 1674. For a detailed discussion of the reforms, see Hunter and Wood 'Towards Solomon's house' (note 6, esp. pp. 190–2). The authors take Petty's return from Ireland in August 1673 to be the main spur, but the entry in Hooke's *Diary* for 22 August 1673 suggests that it was Hooke who stimulated Petty. (Hooke had spoken about the problem with Hoskins as early as June.) Petty was elected to the Council in November 1673, but was not present on 18 June 1674 when, judging by Hooke's *Diary*, the matter was first raised. Petty did preside at the Council meetings in the autumn (27 August, 29 September, 15 and 19 October) and was present at those of 7 October and 9 November when Brouncker presided. Much of the Council's attention at these meetings was devoted to finance and the question of how best to persuade Fellows to pay their arrears.

13 Printed by Hunter and Wood, 'Towards Solomon's house' (note 6) as Document B, from Royal Society MS. MM 4.72. The writer was a Council

member who (unlike Petty) had *not* been asked for his views; Needham was present at the Council meeting of 7 October when Petty was asked to draw up a proposal, and he knew of the plans for having lectures, but was not present at later Council meetings. Michael Hunter (private communication) prefers Croone as a possible author.

14 For the details, see Oldenburg, *Correspondence*, vols. xi and xii.

15 For the meetings at Joes Coffee House (later the Mitre Tavern) see Hooke's *Diary* entries of 10 December and 11 December 1675.

16 Hooke notes that he was to 'speak to Sir J. More, Mr Wild, Mr Hill, Mr Hoskins, Mr Henshaw', while Wren had brought in Dr Holder. At the first meeting on the next day Hooke 'shewd that Mr Newton had taken my hypothesis of the puls or wave' for his theory of the transmission of light.

17 Entry for 2 July. The reason for the name is not obvious. See, further, entries for 8, 13, 14, 15 and 31 July to 10 August.

18 Hunter and Wood, 'Towards Solomon's house' (note 6, pp. 199ff., and for their dating, pp. 242–4); it could have been written as late as 1700. One version is there printed as Document C, pp. 232–9, taken from Royal Society Classified Papers xx.50, fos. 92–4. It is without title, but had been endorsed 'Proposals for advancement of the R.S.'. There is no mention of it in the surviving pages of Hooke's *Diary*.

19 Cf. Hunter and Wood, 'Towards Solomon's house' (note 6, p. 203), with a summary of Waller's reaction to Sloane's account. See also M. Hunter, *The Royal Society and its Fellows 1660–1700*, pp. 45–7. The following list of names in the text is taken directly from the Journal Book copy.

20 Royal Society MS. DM 5.12. It mentions the Berlin Academy, founded in 1700, and also refers to 'his Present Majesty', who must be either William III (Queen Mary died in 1694) or just possibly George I, who came to the throne in 1714. For its ascription to Hooke (which would date it to 1701–3), see Hunter and Wood, 'Towards Solomon's house' (note 6); they take it to have a 'similarity in appearance' to the (undateable) proposal which *is* clearly by Hooke, already discussed. (See M. Hunter, *Establishing the New Science*, note ii, p. 244.) The document was partly printed by Bluhm, 'Remarks on the Royal Society's Finances 1660–1768', p. 92; Bluhm rightly commented upon the notion of the Royal Society as a research foundation. In the nineteenth century Babbage and others contrasted the last two institutions (of Paris and London) in much the same way.

21 Westfall, *Never at Rest*, p. 632, writes, 'he came to the presidency armed with a "Scheme" ...' without giving any reason for his dating, although (p. 685) he also notes that the Council minutes for 30 March 1713 record that after Hauksbee's death Newton proposed four committees 'for the more effectual promoting of the Ends of the institution of this Society'.

The document is printed in entirety in Brewster, *The Life of Sir Isaac Newton*, vol. i, pp. 102–4; he regarded it as conforming perfectly to the needs of the Royal Society of the 1830s! The 'Scheme' exists in half a dozen drafts; I have used that of Cambridge University Library MS. Add. 4005, no. 2.

22 As by Heilbron, *Physics*.

23 As by Shapin, especially in 'The house of experiment in seventeenth century England'.

24 The difficulties encountered in accepting accounts of Newton's optical experiments and of repeating them are discussed in Schaffer, 'Glass works'. Schaffer would not altogether agree with me that readers were becoming more sophisticated at this time.

CHAPTER 3

1 The minutes for the period 1660–87 are reproduced with reasonable accuracy (except for spelling) in Birch, *History*. Birch gives useful references to the Royal Society's Letter Books (which contain copies of a majority of incoming letters and some outgoing ones) and sometimes prints material read at meetings. In a very few cases Birch records no meeting because the Journal books are incomplete. Occasionally rough minutes exist for these meetings and have been reproduced by Kaye, 'Unrecorded early meetings of the Royal Society'; in other cases Hooke's *Diary* often records the fact of the meeting and something about its content. All accounts of early meetings, with these exceptions, are taken from Birch; later ones are taken from the Journal Book copies (made in the eighteenth century for the use of the Presidents) unless otherwise specified.

2 Wallis wrote two accounts, one in 1678 in his *A Defence of the Royal Society* and one in 1696/7, for which see Scriba, 'The autobiography of John Wallis F.R.S.', esp. pp. 39–40.

3 Adamson, 'The Royal Society and Gresham College 1660–1711', is authoritative.

4 Lyons, *The Royal Society 1660–1940*, p. 37.

5 Birch, *History*, vol. i, pp. 50–1.

6 Cf. Hooke's letter to Boyle of July 1663 printed in Birch, *Boyle*, vol. vi, p. 486. In 1664 Hooke also became Cutlerian Lecturer; see M. Hunter, *Establishing the New Science*, pp. 279–336.

7 The Council minutes of 3 June 1663 declare that 'Mr Hooke was elected a Fellow of the Society by the Council and exempted from all Charges', but presumably this is an error and only the last clause is significant.

8 See Birch, *History*, vol. ii, p. 60, for 28 June 1665.

9 Birch, *Boyle*, vol. vi, p. 502, reprinted in Gunther (ed.), *Early Science in Oxford*, vol. vi, p. 248.

10 See Birch, *History*, vol. i, pp. 249–50.

11 This must be related to Power's work as described in Book 3 of his *Experimental Philosophy* of 1664, although his name is not mentioned.

12 Lower to Boyle, 18 January 1661 (*sic*, presumably 1661/2) printed in Birch, *Boyle*, vol. vi, p. 462.

13 *Phil. Trans.* iii, no. 35, 18 May 1668, pp. 672–82.

14 Boyle's letter, with Lower's own account, was published in Lower, *De Corde*, pp. 178–85. See *Phil. Trans.*, i, no. 20, 17 December 1666, pp. 352–7, and for the English translation of *De Corde* by K. J. Franklin, see Gunther (ed.), *Early Science in Oxford*, vol. ix.

15 Oldenburg, *Correspondence*, vol. iii, p. 235, letter to Boyle of 25 September 1666.

16 Cf. false *Phil. Trans.*, no. 27, dated 12 July 1667, and repudiated by Oldenburg in true no. 27 of 23 September 1667. (Oldenburg was in the Tower during July, falsely accused of treasonable activities.) The account by Denis was produced on 14 July. See Hall and Hall, 'The first human blood transfusion', correcting on a number of points A. D. Farr, 'The first human blood transfusion', *Medical History*, **24**, (1980), 143–62.

17 Boyle, Hooke and Lower, while readily performing animal experiments, all sometimes expressed compunction. The latest discussion of this fact (sometimes overlooked) is Maehle, 'Literary responses to animal experimentation', esp. p. 32.

18 Now marked by Arundel Street on the East Side of King's College.

19 Hunter and Wood, 'Towards Solomon's house', relates the proposal quoted below to financial problems, the collapse of committees and the general structure of the Society; see M. Hunter, *Establishing the New Science*, p. 188.

20 For Balle, see Armitage, 'William Ball, F.R.S. 1627–90'.

21 On 2 January; he produced the model on 16 January.

22 On 6 February.

23 Fortunately, because the Society could not afford to disburse what amounted to half of any Fellow's annual subscription, especially as many were gravely delinquent in payment. Hunter and Wood's statement ('Towards Solomon's house', p. 52/p. 189) that such a medal 'was offered' is presumably a slip of the word processor. The proposal does foreshadow the institution of the Copley Medal in the eighteenth century (see below, Chapter 7), but now, although it was further discussed by the Council on 13 April, the necessary money was not forthcoming.

24 See Oldenburg, *Correspondence*, vol. v, p. 103 (26 October 1668), letter to

Huygens, and p. 117 (29 October 1668), letter to Wren. Both produced papers, later published in *Phil. Trans.*

25 That of Huygens was published in *Phil. Trans.*, iv, no. 46, 12 April 1669; of Wallis, **iii**, no. 43, 11 January 1668/9; and of Wren, *ibid.*, pp. 867–8. Huygens had received Wren's account by 6 February 1669 (N.S.).

26 For Wallis's report see Oldenburg, *Correspondence*, vol. vii, pp. 559–64, letter of 7 April 1671, and for Oldenburg's letter of 14 April, see pp. 570–2.

27 Royal Society MS. DM 5.11 printed by Hunter and Wood, 'Towards Solomon's house', pp. 78–80/pp.223–35, and discussed by them at some length.

28 For example, Hunter and Wood, 'Towards Solomons' house', p. 53/p. 190.

29 See M. Hunter, 'Early problems in professionalizing scientific research: N. Grew (1641–1712) and the Royal Society', *Notes & Records*, **36** (1982), 189–209, esp. pp. 191–3, reprinted in M. Hunter, *Establishing the New Science*, pp. 261–78. Grew was sponsored by Wilkins and financed by contributions from various Fellows.

30 Newton, *Correspondence*, vol. i, p. 73; Oldenburg, *Correspondence*, vol. viii, pp. 468–73; and *Oeuvres Complètes de Christiaan Huygens*, vol. vii, pp. 128–31.

31 In his letter of 18 January, Newton, *Correspondence*, vol. i, pp. 82–3, summarized in Oldenburg, *Correspondence*, vol. viii, p. 482.

32 On 14 March 1671/2. Boyle had discussed the colours on soap bubbles in his *Experiments and Considerations touching Colours* of 1664.

33 See p. 33 above, under 1663.

34 *Nove Experimenta (ut vocantur) Magdeburgica de Vacuo Spatio* (Amsterdam, 1672).

35 27 November 1672.

36 For Lister, see meetings of 18 January, 29 February, 1 May, 30 November and 4 December, when his letter on plant anatomy was given to Grew for comment, duly given on 11 December.

37 Cf. the meeting of 22 January 1672/3. Hunt (d. 1713) joined Hooke on 9 January 1672/3 as the Society's employee, in whose service he spent the remainder of his life.

38 For Hooke see, e.g., meetings of 5 February, 5, 19 and 26 March, 2 and 9 April, 14 and 28 May, and 4 June. For Boyle see especially 22 and 19 January, 14 May, and 13 June.

39 See, e.g., meetings of 15 and 22 January 1673/4; his Cutlerian Lecture was 'Animadversions of the first part of *Machina Coelestis* ... of Hevelius' (London, 1674). For magnetic experiments, see 12 and 19 February. The Gregorian telescope was first described to the Society on 26 March 1673 by the reading of a letter from Gregory to Collins; see Birch, *History*, vol. ii, pp. 79–82.

40 See the minutes of the meetings on 5 and 12 February.

CHAPTER 4

1 'To the Reader', *New Experiments Physico-Mechanical*, in Birch, *Boyle*, vol. i, p. 1.

2 As with the reference to the fact that work had been done on watches and the possibility that springs could be used to regulate them, a suggestion made by Hooke, without any details. Unfortunately, because this was in large part the basis for Hooke's claim in 1675 that his ideas had been later 'betrayed' to Huygens in correspondence.

3 For the Latin text and English translation see *The Record of the Royal Society*, pp. 215–37. The 1663 Statutes, the first enacted, are pp. 287–301; those concerning experiments are in Chapter V, and the duties of the Secretaries are defined in Chapter X.

4 A. Hill, *Familiar Letters*, pp. 107–11, letter of May 1663: 'several discourses offered to the Society on forest trees have been recommended to Mr Evelyn to compile into one booke, which is now in the press, and is the first published by order.'

5 Witte was a great admirer of English learning and had taught himself English in order to read English books, with which he begged Hill to supply him.

6 He had been proposed on 8 May 1661; but was elected only on 26 February 1662. He was not an original Fellow but was re-elected on 1 July 1663. Croone seems to have learned of Power's interests through Richard Towneley, a virtuoso, experimenter and gentleman.

7 See Oldenburg, *Correspondence, passim*; also M. B. Hall, 'Oldenburg and the art of scientific communication', and 'The Royal Society's rôle in the diffusion of information in the seventeenth century'.

8 BM MS Add 4441, f. 27 (contractions spelled out). His petition was discussed at a Council meeting of 27 April 1668 and resulted in his receiving fifty pounds, and a year later he was promised a regular salary of forty pounds a year. Weld, *History of the Royal Society*, vol. i, p. 135, note 12, gives a slightly different version, with the figure of '50 persons', and, not having noted the relevant Council minute, dates the document to 1664.

9 Glanvill, *Plus Ultra*, p. 103.

10 Oldenburg, *Correspondence*, vol. vii, pp. 110–14 and 107–9; it was in reply to a letter from von Boineburg asking Oldenburg to act as postal agent for a letter to Prince Rupert and introducing the young Leibniz, who also sent letters to Oldenburg and to Hobbes (Oldenburg, *Correspondence*, vol. xiii, pp. 421–3). The touch of German pride in Oldenburg's reply reflects the

opening sentence of von Boineburg's original letter: 'How greatly the Royal Society esteems you, who are a German, on account of your merits, is surely not unknown to us.' Normally, Oldenburg showed a distinctly English bias.

11 Oldenburg, *Correspondence*, vols. ii and iii.

12 Oldenburg, *Correspondence*, vol. x.

13 Oldenburg, *Correspondence*, vols. ix–xiii; when Line died in 1675 the controversy was continued by his pupil, Anthony Lucas.

14 Newton, *Correspondence*, vol. i, pp. 247, 252 and 294, of which there are extracts in Oldenburg, *Correspondence*, vol. ix, 427–30, and vol. x, p. 44.

15 Newton, *Correspondence*, vol. ii, letters to Hooke of 1677–8.

16 The works edited by Oldenburg were *Dissertatio epistolica de bombyce* (1669), for which the order was given by the Council on 22 February 1668/9, *Dissertatio epistolica de formatione pulli in ovo* (1673), both, as their titles imply, mainly composed of letters exchanged between Malpighi and Oldenburg, and *Anatomes plantarum idea* and *Anatomes plantarum pars prima* (1675), these last two published together.

17 Each issue was dated, the date presumably that on which it was passed to the printer. It was intended that there should be twelve issues or numbers in each volume, but the pressure of external events – plague, fire, Oldenburg's imprisonment in the summer of 1667, printers' holidays – made their appearance irregular, especially in the 1660s. In the end, there were to be 142 issues in the twelve volumes, 1665–79, Oldenburg's last having been no. 136 of June 1677.

18 See Oldenburg, *Correspondence*, vol. ii, e.g. pp. 588, 590–1 and 646–8.

19 See Oldenburg, *Correspondence*, vol. ix, 517, note 3, and *Phil. Trans.*, **vi**, no. 75 for 18 September 1675, pp. 2269–70. Oldenburg, *Correspondence*, vol. ix contains letters from the translator, Christoph Sand. There were many irritating errors, but perhaps not so many as in the volume for 1666, translated by John Sterpin, whose exculpatory letter is in Oldenburg, *Correspondence*, vol. vii, pp. 468–9. Sand had sought Oldenburg's advice frequently.

20 Number for 30 March 1665 (N.S.); the relevant words are 'On a pris le soin d'y [in England] faire un Journal en Anglais sous le titre de Philosophical Transactions, pour faire sçavoir à tout le monde ce qui se découvre de nouveau dans la Philosophie.'

21 Oldenburg, *Correspondence*, vol. vi, p. 259, letter of 5 October 1669.

22 Oldenburg, *Correspondence*, vol. iii, pp. xxvii–xxix, 448–53 and 471–2.

CHAPTER 5

1 On 17 December Hooke proposed to the Council the undertaking of a series of magnetic experiments 'that would require an apparatus of instruments for the making of them' and asked the Council to order their preparation; cautiously, the Council demanded details of the experiments first. In 1674 Hooke was also, as later, reading his Cutlerian Lectures; see M. Hunter, *Establishing the New Science*, chap. 9, 'Science, technology and patronage', pp. 279–338.

2 16 December 1675, presumably performed by Hooke, since he had promised 'to prepare it' for the meeting. It succeeded on 13 January 1675/6. There was more repeating and testing of Newton's experiments at the meetings of 2 March 1675/6 and 27 April 1676.

3 See Chapter 2.

4 Daniel Whistler (b. 1619, F.R.S. 1661), physician and a Gresham Professor, often previously on the Council as at that time. The most convenient source of biographical information is M. Hunter, *The Royal Society and its Fellows 1660–1700*.

5 Philip Packer (b. *c.* 1620), a barrister; he was to serve on the Council in 1679 and in the 1680s.

6 Summarised by Hooke in his *Diary* as 'Grew take notes also, but I to draw them up'.

7 On 1 Nobember and 6 December; for Hill's paper, see Chapter 9.

8 16 and 23 January, 6, 13 and 20 February.

9 On 20 March 1678/9.

10 No doubt promising that Papin's appointment would increase foreign correspondence which had languished in the past two years. Since it was only agreed to give Papin either eighteen pence or two shillings per letter copied, depending upon length, it is a little surprising that he accepted the employment. Possibly Boyle, who continued to act as his patron, encouraged him to do so. In 1680 there was published *Experimentorum Novorum Physico-Mechanicorum Continua Secunda*, the fruit of four years' collaboration between Papin and Boyle, which must have been written up by Papin; the English version appeared two years later. Boyle is the titular author and presumably devised the experiments to be performed by Papin, who is not named directly.

11 Council minutes for 30 September 1679.

12 But his *Diary*, in recording the presence (or absence) of auditors at his Cutlerian and Gresham Lectures, often at this time records the presence of what he called 'a spy'.

13 Published as 'Lectures of light' by Richard Waller in *The Posthumous Works of Robert Hooke*.

14 The meeting of 3 May, with experiments on coloured solutions derived from Boyle's *Experimental History of Colours*, and the meeting of 26 July for some optical illusions.

15 See *New Experiments ... Touching the Weight and Spring of the Air*, First Continuation, in Birch, *Boyle*, vol. iii, p. 256.

16 The minutes record that MM. Justell and Auzout 'desired, that ... Hubin ... might be admitted', presumably in a letter of introduction. He is described as 'enameller to the French King' and was a well-known instrument maker. Justel had emigrated to England in 1682.

17 *Phil. Trans.*, **xii**, no. 148, 1683.

18 See Birch, *History*, vol. iv, p. 324; on 17 September 1684 Boyle, to whom the copy sent to the Society was given, sent in a Latin abstract, while Slare reported on it in English.

19 10 and 17 February 1685/6. He compared 'a direction by the naked eye with a radius of ten feet, with that of a telescope of eight inches'.

20 One of the first of these discourses to be so designated.

21 12 January 1686/7.

22 See Newton, *Correspondence*, vol. iii.

23 Lord Vaughan, now Earl of Carbery (P.R.S. 1686–9), was a minor mathematician, as was the Earl of Pembroke (P.R.S. 1689–90), who was also an antiquary; Southwell (P.R.S. 1690–5) had been a virtuoso in his youth but was now mainly a man of affairs; Montagu (P.R.S. 1696–8), later Earl of Halifax, Newton's patron, was a Member of Parliament and a Treasury official; and Somers (P.R.S. 1698–1703) was a lawyer and politician.

24 See Gunther, (ed.), *Early Science in Oxford*, vol. x.

25 Halley's activities from 1688 to 1698 are conveniently listed as extracts from the minutes of the Society's meetings in MacPike, *Correspondence and Papers of Edmond Halley*, pp. 210–40; MacPike also gives relevant extracts (pp. 182–6) of Hooke's 1688–93 *Diary*.

26 Hooke's lists never contain the names of more than sixteen persons besides himself, and usually fewer. Possibly he sometimes omitted or simply forgot names of inactive Fellows or even those who took no part in the social meeting at Jonathan's coffee house on the same day.

27 Picards' experiment was discussed at the Royal Society meeting of 25 May 1676. The flash of light in both experiments is, of course, an example of an electric discharge through a rarefied gas, here air. The Mr Hill who spoke is probably Abraham Hill, not Oliver Hill. The phenomenon was not pursued until the next century, when it was taken up first by Johann Bernoulli (1700) and, a little later, by Francis Hauksbee.

28 The paper had been given to Aston, then Secretary, in October 1683; it was discussed at the meetings of 17 January and 9 February 1691/2.

29 Published in Hooke's *Posthumous Works*.

30 As shown in the minutes; see below, note 32.

31 Etienne François Geoffroy (1672–1731) came to England in 1698 as physician to the French Ambassador and was soon elected F.R.S. For his later relations with the Society, see Chapter 6.

32 Best known for his abridgement of the first twenty-one volumes (to 1700) of *Phil. Trans.*, approved by the Council in 1703, published 1705. He was proposed a Fellow in May 1699 but not elected until 1702. Evidently strangers were currently being admitted fairly frequently. When Derham found some of Lowthorp's 'observations' in Hooke's papers, he published them in Hooke's *Philosophical Experiments* of 1726; these were subsequently published as by Hooke in Gunther (ed.), *Early Science in Oxford*, vol. vii, p. 786, along with Derham's note correctly ascribing them to Lowthorp.

33 It is not possible to determine whether this was William Bridgeman (F.R.S. 1679) or his son Orlando (F.R.S. 1696), but perhaps more probably the latter.

34 See Adamson, 'The Royal Society and Gresham College' 1660–1711'.

CHAPTER 6

1 The Council met on 13, 24 and 27 September, but although it was then formally adjourned until 11 October, there are no further Council minutes until late November.

2 Leeuwenhoek had written to Oldenburg on 5 October (N.S.), unaware of his death. Now, eleven days later, he wrote to Brouncker as President, enclosing a paper of observations; which observations he did not say but almost certainly those printed in *Phil. Trans.*, **xii**, no. 142 (December/February 1678/9) in a Latin translation sent by the author, together with testimonials of eye-witnesses to the truth of his findings. See Birch, *History*, vol. iii, p. 347, 1 November 1677, and *Alle de Brieven*, vol. ii, pp. 236–99. Brouncker had passed the letter to Henshaw, Secretary and Vice-President, to whom he had resigned all business during this period.

3 According to Leeuwenhoek's reply of 4/14 January 1678, less than a week later; see *Alle de Brieven*, vol. ii, p. 305.

4 Newton, *Correspondence*, vol. ii, pp. 239, 253, 264, 265 and 269. In 1678 Aubrey was in correspondence with Lucas on his own account; see M. Hunter, *John Aubrey and the Realm of Learning* (London, 1975), p. 62, note 9. Hunter has no occasion to mention Aubrey's 1677 correspondence with Newton and Lucas.

5 Justel wrote to Leibniz on 17/27 September 1677 in very regretful tones (*Sämtliche Schriften*, (Leipzig and Berlin, 1923 ff.), vol. **i**, ii, pp. 293–4), and

James Crawford (English resident in Venice) informed Malpighi on 24 September from London (Adelmann, *Correspondence of Marcello Malpighi*, pp. 759–60).

6 For the letter sent to Huygens see his *Oeuvres Complètes*, vol. viii, p. 66. Grew's letter to Williamson, 11 April, is P.R.O. MS 29/403, no. 19. There is no trace of such a letter in Newton's MSS.

7 Royal Society MS 243, Letter 45 to Towneley, 13 February.

8 See Gunther, (ed.), *Early Science in Oxford*, vol. iv, for the minutes from October 1683 to 1690. The meetings were infrequent after 1686.

9 See Hoppen, *The Common Scientist in the Seventeenth Century*.

10 Cf. Royal Society Letter Book VIII for the years 1677 to 1682/3.

11 See Halley to Leeuwenhoek, 2 March 1685/6, in MacPike, *Correspondence and Papers of Edmond Halley*, pp. 56–7, from the copy in Royal Society Letter Book IX, also p. 66, and *Alle de Brieven*, vols. v, vii and viii.

12 One such was Francis Lodwick (F.R.S. 1681), who in 1687 was given some of Leeuwenhoek's letters to translate.

13 Cf. Lyons, 'Biographical Notes'. Aston recovered his temper to serve on the Council in the later 1690s. Cf. Birch, *History*, vol. iv, p. 449 and note p. 451, and Weld, *History of the Royal Society*, vol. i, pp. 302–5, all quoting Halley's letter to Molyneux of 27 March 1685. For Aston's side see his letter of 10 December to Musgrave in Gunther (ed.), *Early Science in Oxford*, vol. xii, p. 105. Writing to John Caswell in Oxford (he had sent in a paper read to the Society), on 9 July 1686, Halley apologised for the interruption to the correspondence between the Royal and the Oxford Societies, attributing it to 'the misunderstandings of some, and ill will of others'. (Gunther (ed.), *Early Science in Oxford*, vol. xii, p. 111.)

14 Robinson (F.R.S. 1684), a friend of Ray (see Raven, *John Ray, Naturalist*, pp. 207–8) was elected Secretary on 30 November 1685; as a friend of Aston he resigned (9 December) when Aston did.

15 Letter Book XI, Part 2, headed 'Letter ... desiring a Correspondence' but bearing no date. Clayton's first paper is dated 12 May 1688; his second is not dated in its published form.

16 Robinson, 'The administrative staff of the Royal Society 1663–1861', p. 193.

17 Quoted in Lyons, 'The Officers of the Society 1662–1860', pp. 122–3. Lyons found a number of similar letters 'in the Letter Books'.

18 *A Journey to Paris in the Year 1698* (London, 1699), p. 97.

19 Cf. Jacquot, 'Sir Hans Sloane and French Men of Science'.

20 Robinson, 'The administrative staff of the Royal Society, 1663–1861'; Adelmann, *Marcello Malpighi and the Evolution of Embryology*; and Journal Book for 7 November 1688, 3 May 1696 and 28 October 1696.

21 He had previously (1676–8) published and illustrated Virgil (manuscript

in the British Library). The best recent account is by Ezell, 'Richard Waller, S.R.S'; this supersedes all others.

22 Lister's letter of 11 February 1681/2 was read to the Society on 15 February; the members present then decided to encourage the publication and did so at a meeting on 5 April. But it was not until 11 March 1684/5 that Tancred Robinson could produce the prepared manuscript. The illustrations made the book so expensive that the Society agreed to underwrite the cost, with the help of contributions from some Fellows (a guinea a plate). See Birch, *History*, vol. iv, for 25 March, 27 May and 11 November 1685, on which latter date Lister, as Vice-President, licensed it. Ray had received copies by 5 May 1686.

23 Thus Sloane had promoted an English translation (by Nicholas Staphorst, his former teacher of chemistry) published in 1683; see Brooks, *Sir Hans Sloane the Great Collector and his Circle*, p. 123.

24 *Diary* for 5 and 9 October.

25 He wrote to John Beale (6 March 1680) that he favoured a summary of the minutes, circulation to be restricted to Fellows, plus a fortnightly journal (letter quoted in M. Hunter, *Establishing the New Science*, p. 199). On 9 November Tyson told Plot (in Oxford) that he promised 'a Collection out this week, and so once a fortnight or a month', (Gunther (ed.), *Early Science in Oxford*, vol. xii, p. 8).

26 *Phil. Trans.*, **xiii**, no. 143 (January 1682/3), p. 2. M. Hunter, *The Royal Society and its Fellows 1600–1700*, note 2, pp. 141–2, drawing on accounts of payments 1683–92, adds details of the divided editorial responsibility.

27 'Advertisement', *Phil. Trans.*, **xvii**, no. 193 (1691).

28 Council minutes, confirmed by the entry for that date in Hooke's *Diary* (Gunther, (ed.), *Early Science in Oxford*, vol. x).

29 This may be the number whose first printed sheets Hooke saw on 18 March 1692/3, although he called it no. 195, which in fact was dated 1692, almost certainly however Old Style, making it really 1693. Hooke received his copy of one of these numbers on 29 March 1693.

30 See Ezell, 'Richard Waller, S.R.S.'. Note 20, pp. 223–4.

31 He mentioned it on 5 July but corrected the last sheet on 26 July.

CHAPTER 7

1 Newton had been elected to the Council in 1697 and 1699 but never attended Council meetings; he seems to have attended very few meetings of the Society until his Presidency. (Westfall, *Never at Rest*, lists occasions in 1675, 1689 and 1699 only.)

2 See Guerlac, 'Sir Isaac and the Ingenious Mr. Hauksbee', 'Francis
  Hauksbee', and 'Francis Hauksbee: Expérimenteur au Profit de Newton'.
3 By Guerlac; see 'Sir Isaac and the Ingenious Mr. Hauksbee', pp. 240–1.
4 See Heilbron, *Physics*.
5 It is worth recalling that teratology is now a respectable branch of medical
  science, although until recently the frequent descriptions given in the
  seventeenth and eighteenth centuries of two-headed calves and Siamese
  twins were seen as being laughably quaint.
6 They were usually printed in the *Phil. Trans.*, except as already noted
  during those years when Halley was editor, namely 1686–7 and 1715–19.
7 Especially by Vallisnieri (F.R.S. 1703) from Italy and Tournefort from
  France.
8 Copley's legacy was not regularly disbursed until 1737, when Martin
  Folkes (then a Vice-President) suggested its conversion to an annual prize
  with a medal, as still awarded. Gray received the medal for 1731 and
  1732, and was in the latter year finally elected a Fellow; Desaguliers
  received the medal for 1734, 1736 and 1741. It soon became a prize for
  experimental contributions 'published or communicated' to the Society, no
  longer specifically *exhibited* at meetings.
9 Villette's powerful burning glass had been known to the Society since 1665
  and accounts of it were published in *Phil. Trans.*
10 Presumably not John Godfrey (F.R.S. 1715), but rather Ambrose, whose
  papers were later published in *Phil. Trans.*
11 Possibly John Brown, two of whose papers on Epsom salts were published
  in *Phil. Trans.* in 1722–3, who was also the author of an experimental
  paper describing Prussian blue as well as one on camphor, published in
  Phil. Trans. in 1724–5.
12 By Heilbron, who being only concerned with physics does not consider the
  empirico-experimental aspects of non-physical science.
13 An earlier paper on mineral water by Mr Cay was published in *Phil.
  Trans.* for 1698.
14 It was to be made by Francis Hauksbee the younger, nephew and
  associate of the former Curator of Experiments, also an instrument maker,
  who had recently been appointed Clerk and Housekeeper to the Society.
  John Horsley (F.R.S. 1727), one of whose papers was published in this
  year's *Phil. Trans.*, was a Northumberland clergyman and archeologist.
15 Teste John Byrom (F.R.S. 1724), who had studied medicine but was never
  to practice, making his living by teaching shorthand. See Talon (ed.),
  Selections from the Journals & Papers of John Byrom, p. 71 (11 March 1724).
16 On 18 February 1725 he showed a fetal membrane, comparing it with the
  description published by Dr. Richard Hale in *Phil. Trans.* 'to Justify the
  Truth of the Description'.

17 They were to show the effects of centrifugal force in the rotation of a spheroid. This was described by Byrom (note 15 above) as being 'about the leaden ball's cohesion'; and this experiment (the cohesion of balls of different materials) was also performed later (1727).

18 Teste Byrom, for 11 February 1730/31.

19 In eight years (1761, 1762, 1763, 1765, 1774, 1779, 1790 and 1793) no award was made, but three medals were awarded in 1766 and two in 1791.

20 The minutes of 16 March 1731/2 record Mr Godfrey's performance of Dr Frobenius's experiment before the Society (on phosphorus). For 11 March 1736 see Byrom (note 15 above), who described the experiments as consisting 'of a little bit of cork suspended by a thread and turning round a ball of iron upon a cake of resin from west to east'.

21 See Journal Book for 27 March 1735. A complaint that 'members convene together to discourse of various points in such a manner as to draw off the attention of the Society from the business they are upon' is recorded; this it is noted is contrary to the Statutes, which specify that members should always address the chair.

CHAPTER 8

1 See A. R. Hall, *Philosophers at War*.

2 Cohen, 'Isaac Newton, Hans Sloane and the Académie Royale des Sciences', esp. pp. 102–16, and Guerlac, *Newton on the Continent*, pp. 102–3.

3 As already noted, the content of experimental and observational physics has been extensively and statistically analysed by Heilbron, *Physics*, and there is no need to repeat his study here. But once again it should be recalled that he was not at all concerned with other branches of natural philosophy.

4 Cohen, 'Isaac Newton, Hans Sloane and the Académie Royale des Sciences', note 2, p. 96.

5 Newton, *Correspondence*, vol. vi, p. 145 (9 June 1714, N.S.).

6 See, however, the excellent article by Casini, 'Les débuts du Newtonianisme en Italie, 1700–1740', detailed and authoritative, to which I am much indebted, and also M. B. Hall, 'The Royal Society and Italy 1667–1795', *Notes & Records*, **37** (1982), 63–81.

7 Cf. Newton, *Correspondence*, vol. vi, pp. 506–7 (December 1707), and vol. vii, p. 483 (May 1718).

8 Newton, *Correspondence*, vol. vii, pp. 278 and 330.

9 For what follows, see Casini, 'Les débuts du Newtonianisme en Italie, 1700–1740', *passim*.

CHAPTER 9

1 The Society judged each potential Fellow on his merits, which did not necessarily include contributions to natural philosophy. The term 'virtuoso' covered a range so wide that a poet like Dryden or a courtier with no pretense to anything other than a vague interest or a lawyer with an inclination towards natural philosophy were all potentially acceptable, provided that they were willing to agree to pay their annual dues, in itself a valuable contribution.

2 Cf. Oldenburg, *Correspondence*, vols. v and vi, *passim*.

3 Pepys, *Diary*, 14 April 1664, 2 January 1664/5, 1 March 1664/5, 18 April 1667.

4 The evidence is John Evelyn's *Diary* for 13 December 1685. Slare had done much work on phosphorus.

5 A. Hill, *Familiar Letters*, to John Brooks.

6 Like Ralph Thoresby (F.R.S. 1697), who was thrilled to be elected and to attend meetings, cf. J. Hunter, *The Diary of Ralph Thoresby*, vol. i, *passim*; or John Byrom (F.R.S. 1724), who attended meetings in the 1720s, cf. Talon, *Selections from the Journals & Papers of John Byrom*.

7 See Cavazza, 'Bologna and the Royal Society in the seventeenth century'.

8 Cavazza, 'Bologna and the Royal Society in the seventeenth century', p. 114.

9 See Oldenburg, *Correspondence*, vols. v *et seq.*

10 Oldenburg, *Correspondence*, vol. vi, pp. 421 ff.

11 Malpighi wrote about both; see Oldenburg, *Correspondence*, vol. iv, p. 272. For Cornelio and tarantulas, see, esp., Oldenburg, *Correspondence*, vol. viii.

12 See Cavazza, 'Bologna and the Royal Society in the seventeenth century', pp. 109ff.

13 See Westfall, *Never at Rest*, p. 796, and Casini, 'Les débuts du Newtonianisme en Italie, 1700–1740'.

14 Huygens, *Oeuvres Complètes*, vol. viii, p. 66.

15 Birch, *History*, vol. iii, p. 10.

16 See Oldenburg, *Correspondence*, vol. x, pp. 315–16.

17 His correspondence with Abraham Hill is printed in A. Hill, *Familiar Letters*; the first letter is in French, the later ones (the correspondence continued until 1669) in Latin.

18 This was a society of physicians formed solely to publish the research papers of its members and held no meetings. It continued into the twentieth century.

19 Quoted in Ornstein, *The Rôle of Scientific Societies in the Seventeenth Century*, pp. 175–7. Sturm's own account is in his *Collegium Experimentale sive Curiosorum, in quo primaria hujus Seculi Inventa & Experimenta*

Physico-Mathematica A. 1672 quibus Naturae Scrutoribus spectanda exhibuit et ad causas suas naturales demonstrata methodo reduxit (Nuremberg, 1676), whose title clearly states the intended purpose of the Society.

20 This Oldenburg did a week later.

21 See Oldenburg, *Correspondence*, vol. x, pp. 432–3, letter of 12 January 1673/4.

22 Birch, *History*, vol. ii, p. 482, meeting of 25 May 1671.

23 Wallis's full report (in Latin) dated 7 April 1671 was printed in *Phil. Trans.*, **vi** (1671), pp. 2227–30; see also Oldenburg, *Correspondence*, vol. vii, pp. 559–65.

24 See A. R. Hall, *Philosophers at War*.

25 See the minutes of the meetings of 18 March 1679/80, 9 March 1680/1, 1 November 1682, 26 November 1684. He also sent a comment on Halley's magnetic hypothesis.

26 Cf. *Olai Borrichio Itinerarium 1660–65*, vol. iv, p. 201 (8 January 1665).

27 Oldenburg, *Correspondence*, vol. viii, pp. 281 and 283; cf. Oldenburg's earlier letter, *ibid.*, pp. 191–3.

28 According to Hooke's *Diary*, entries for 14 and 15 May 1679 and 25 September 1679. The letters between Roemer and Flamsteed are in Royal Greenwich Observatory MSS. RR D42, ff. 8–12, now in the Cambridge University Library. See also M. B. Hall, 'Roemer et l'Angleterre' in *Roemer et la Vitesse de la Lumière*.

29 Costabel, 'Le registre académique 'Journeaux d'Angleterre' et Mariotte'.

30 Especially the famous 'experimentum crucis', for Mariotte insisted that all rays of light, whatever their colour, were modified in passing through a prism. See his *Traité des couleurs* (E. Mariotte, *Oeuvres*, Leiden 1717, pp. 196–320) and A. R. Hall, 'Mariotte et la Royal Society', pp. 33 ff.

31 Letter of 31 December 1698 (N.S.) (BM Sloane MS. 4025, f. 132, partly printed in Cohen, 'Isaac Newton, Hans Sloane and the Académie Royale des Sciences', pp. 82–3; see also Jacquot, 'Sir Hans Sloane and French men of science'.

32 See Cohen, 'Isaac Newton, Hans Sloane and the Académie Royale des Sciences', p. 83. It is probable that Sloane failed to inform Newton that his nomination carried with it the obligation to send reports of his work to Paris.

33 For Huygens's letter to Leibniz of 4 February 1692, see Huygens, *Oeuvres Complètes*, vol. x, p. 239; for Leibniz's letter of 8 January 1692 see Huygens, *Oeuvres Complètes*, vol. x, pp. 228–9.

34 See A. R. Hall and M. B. Hall, 'Philosophy and natural philosophy: Boyle and Spinoza'. The letters were first published in English by A. Wolf, *The Correspondence of Spinoza* (New York, 1929); for a slightly different English version with full notes, see Oldenburg, *Correspondence*, vols. i and ii.

35 This is a point made by Shapin and Schaffer in *Leviathan and the Air-Pump*.

36 See Schaffer, 'Wallifaction: Thomas Hobbes on school divinity and experimental pneumatics. See also Oldenburg, *Correspondence*, vol. ii, p. 180 and, for 1664 attacks by Wallis on Hobbes's mathematics (ignored by Shapin and Schaffer), Oldenburg, *Correspondence*, vol. iii.

37 Thus Homberg's work on salts leaned heavily on Boyle's work, and Boyle's tests for acids and alkalis were widely adopted. And in the case of Lavoisier's famous quantitative experiment, which showed that water does not turn into earth by long boiling but merely dissolves some of the glass vessel in which it is boiled, it was Boyle who both reported the original experiment (from hearsay) and suggested the method of testing it which Lavoisier followed.

38 Pepys, *Diary*, 1 February 1663/4. Petty, who was present, confirmed this account ten years later in his *Discourse on Duplicate Proportion*, originally read to the Society. Cf. the belief that Charles II, called the Fellows 'his fools' (i.e. court jesters), discussed by Middleton, 'What did Charles II call the Fellows of the Royal Society?'.

39 This is mentioned, and three stanzas of it quoted, by Weld, *History of the Royal Society*, vol. i, note 10, p. 79. It was first published in its entirety by Stimson, 'Ballad of Gresham College', also in her *Scientist and Amateurs*, pp. 57–63, and again by Taylor, from a different source, 'An early satyrical poem on the Royal Society'. There are six MS. copies: BM Add. MSS. 211, 34, 217 and Sloane 1326, and Oxford MS. Ashmole 36 and 37. The work is signed 'W.G.' and in Sloane MS is an anonymous attribution to William Glanville (possibly Evelyn's brother-in-law) but there is no complete agreement as to the identity of the author.

40 William Derham (ed.), *Philosophical Letters between the late Learned Mr Ray And several of his Ingenious Correspondents* (London, 1718), pp. 27–8.

41 Written in the 1660s, perhaps as early as 1664 at about the same time as the second part of *Hudibras*, although only published in Butler's *Genuine Remains*. This stanza is quoted in Stimson, *Scientists and Amateurs*, p. 92. The Fellows themselves were never circle squarers, although they were often taken to be interested in the problem well into the nineteenth century. (There were important discoveries in the rectification of other curves by Wren, Neile and other early Fellows.) It is notable that Wilkins's *Discovery of a World in the Moon* of 1640, with its speculation that flying machines might one day be devised to take man to the moon, was taken as an example of the Society's impracticality, although this was published twenty years before the Society came into existence and was in fact a continuation of a literary tradition as old as the Romans.

42 Literally, 'enlarge upon fleas, lice and themselves', but the pun in the use of 'magnificare' is obvious. Quoted in Glanvill to Oldenburg, 19 July

NOTES TO PP. 164–67

1669; see Oldenburg, *Correspondence*, vol. vi, p. 137. The occasion was the opening of Wren's newly completed Sheldonian Theatre, when South commended Wren for differing from his fellow members of the Royal Society who did such foolish things.

43 Hooke (*Diary*, 1, 2 and 3 June 1676) indicates that he and others took it that he was Shadwell's target.

44 Conclusion, pp. 393–4 of the third edition of 1705.

45 Clearly less influential, because far less successful on the stage, and in any case not really directed against the Royal Society, was Aphra Behn's 1697 farce, *The Emperor in the Moon*, which must have derived directly from Fontenelle's *Entretiens sur la pluralite des mondes* (1686), which she translated into English, her version being published in 1688.

46 The quotation is from *Of Credulity and Incredulity* (1668), as quoted by Syfret, 'Some early critics of the Royal Society', p. 41; here there is also is discussion of Casaubon's *Letter to P. du Moulin* (1669), where he specifically attacks Glanvill and Sprat personally, but exempts the Royal Society. See also Spiller, *'Concerning Natural Experimental Philosophie'*, which reprints the *Letter*.

47 For Steele's remarks in the *Tatler* (no. 236 of 1710) and Addison's in the *Spectator* (no. 262 of 1711) see, conventiently, Stimson, *Scientists and Amateurs*, pp. 127–30.

48 *The Dunciad*, lines 567–70, quoted in Syfret, 'Some early critics of the Royal Society', note 45, p. 50.

Bibliography

Adamson, Ian, 'The Royal Society and Gresham College 1660–1711', *Notes & Records*, **33** (1978), 1–21

Adelmann, H. B., *The Correspondence of Marcello Malpighi*, 6 vols. (Ithaca, N.Y., 1975)

– *Marcello Malpighi and the Evolution of Embryology*, 5 vols. (Ithaca, N.Y., 1966)

Alle de Brieven, see Leeuwenhoek

Andrade, E. N. da C. 'Samuel Pepys and the Royal Society', *Notes & Records*, **18** (1963), 82–93

Armitage, Angus, 'William Ball, F. R. S. 1627–90', *Notes & Records*, **15** (1960), 167–72

Birch, Thomas, *The History of the Royal Society*, 4 vols. (London, 1756–7). [Facsimile in Johnson Reprint series, A. Rupert Hall (ed.) (N.Y. and London, 1968), with the missing minutes reprinted from, Kaye, 'Unrecorded early meetings of the Royal Society']

– *The Life and Work of the Honourable Robert Boyle*, 6 vols. (London, 1772)

Bluhm, R. K., 'Remarks on the Royal Society's finances 1660–1768', *Notes & Records*, **13** (1958), 82–103

Borrichius, Olaus, *Itinerarium 1660–65*, 4 vols. (Copenhagen, 1983)

Brewster, David, *The Life of Sir Isaac Newton*, 2 vols. (Edinburgh, 1850)

Brooks, E. St John, *Sir Hans Sloane the Great Collector and his Circle* (London, 1954)

Brown, Harcourt, *Scientific Organizations in Seventeenth Century France 1620–1680* (Baltimore, 1934)

Butler, Samuel, *Genuine Remains* (London, 1759)

Byrom, John, *see* Talon, Henri

Casini, Paolo, 'Les débuts du Newtonianisme en Italie, 1700–1740', *Dix-huitième Siècle*, **10** (1978), 85–100

Cavazza, Marta, 'Bologna and the Royal Society in the seventeenth century', *Notes & Records*, **35** (1980), 105–23

BIBLIOGRAPHY

– *Setecento Inquieto Alle Origine del'Istituto delle Scienze di Bologna* (Bologna, 1990)

Cohen, I. B., 'Isaac Newton, Hans Sloane and the Académie Royale des Sciences', *Mélanges Alexandre Koyré* (q.v.), vol. I, pp. 61–116

Costabel, P., 'Le registre académique "Journeaux d'Angleterre" et Mariotte', *Mariotte Savant et Philosophe* (q.v.), pp. 321–5

'Espinasse, Margaret, *Robert Hooke* (London, 1956)

Ezell, Margaret A. M., 'Richard Waller, F. R. S. "In Pursuit of Nature" ', *Notes & Records*, **38** (1984), 215–33

Glanvill, Joseph, *Plus Ultra* (London, 1668)

Golinski, Jan, 'A noble spectacle. Phosphorus and the public cultures of science in the early Royal Society', *Isis*, **80** (1989), 11–39

Guerlac, Henry, 'Francis Hauksbee', *Dictionary of Scientific Biography*, vol. VI, pp. 169–75

– *Newton on the Continent* (Ithaca, N.Y., and London, 1981)

–'Sir Isaac and the ingenious Mr. Hauksbee', *Mélanges Alexandre Koyré* (q.v.), vol. I, pp. 228–53

– Francis Hauksbee: Expérimenteur au Profit de Newton', *Essays and Papers in the History of Science* (Baltimore 1977), pp. 107–19

Gunther, R. T. (ed.), *Early Science in Oxford*, vols. IV, VI–XII (Oxford, [printed for the editor], 1925–39). [*See also* Hooke, Lower, Oxford Philosophical Society for details]

Hall, A. Rupert 'Mariotte et la Royal Society', *Mariotte Savant et Philosophe* (q.v.), pp. 33–42

– *Philosophers at War. The quarrel between Newton and Leibniz* (Cambridge, 1980)

Hall, A. Rupert and Marie Boas (eds.), *The Correspondence of Henry Oldenburg*, 13 vols. (1–9, Wisconsin University Press, Madison and London, 1965–73; 10 and 11, Mansell, London, 1975, 1977; 12 and 13, Taylor & Francis, London, 1986)

– 'The first human blood transfusion: priority disputes', *Medical History*, **24** (1980), 461–5

– 'The intellectual origins of the Royal Society – London and Oxford', *Notes & Records*, **23** (1968), 157–68

– 'Philosophy and natural philosophy: Boyle and Spinoza', *Mélanges Alexandre Koyré* (q.v.), vol. II, pp. 241–56

Hall, Marie Boas, *All Scientists Now. The Royal Society in the nineteenth century* (Cambridge, 1984)

– 'La filosofia sperimentale newtoniana e la Royal Society', *Giornale di Fisica*, **30** (1989), 141–50

– 'Oldenburg and the art of scientific communication', *BJHS*, **II** (1965), 277–90

– 'The Royal Society's rôle in the diffusion of information in the seventeenth century', *Notes & Records*, **29** (1975), 173–92

– 'Salomon's house emergent: the Royal Society and cooperative research', in H. Woolf (ed.), *The Analytic Spirit. Essays in the history of science in honor of Henry Guerlac* (Ithaca and London, 1981), pp. 177–94

– 'Science in the early Royal Society', in M. Crosland (ed.), *The Emergence of Science in Western Europe* (London, 1975), pp. 57–77

Heilbron, J. L. *Physics at the Royal Society during Newton's Presidency* (William Andrews Clark Memorial Library, Los Angeles, 1983)

Hill, Abraham, *Familiar Letters which passed between Abraham Hill Esq. . . . and several eminent and ingenious persons*, ed. L. Astley (London, 1767)

Hill, Christopher, 'The intellectual origins of the Royal Society – London or Oxford?', *Notes & Records*, **23** (1968), 144–56

Hooke, Robert, *Lectiones Cutlerianae or a Collection of Lectures* (London, 1679). [Facsimile in R. T. Gunther (ed.), *Early Science in Oxford*, vol. VIII]

– *Diary. See* Robinson and Adams, and Gunther (ed.) 'The Life and work of', vol. X

– 'The Life and Work of', Gunther, (ed.), *Early Science in Oxford*, vols. VI and VII [reprinted from various sources, chiefly the Journal Books of the Royal Society and vol. X, the *Diary 1688–1693*]

– *Philosophical Experiments and Observations of the late Eminent Dr. Robert Hooke, . . . and Other Eminent Virtuoso's in his time*, ed. W. Derham (London, 1726)

– *The Posthumous Works*, ed. Richard Waller (London, 1705; 2nd edn ed. Tom Brown, London 1971)

– (ed.), *Philosophical Collections* (London, 1679–82)

Hoppen, K. T., *The Common Scientist in the Seventeenth Century. A study of the Dublin Philosophical Society 1683–1708* (London, 1970)

Hunter, Joseph (ed.), *The Diary of Ralph Thoresby F.R.S. 1677–1724*, 2 vols. (London, 1730)

Hunter, Michael, *Establishing the New Science. The experience of the Royal Society* (Boydell Press, Woodbridge, 1989). [Note: Because many of the chapters of this book were published in the earlier 1980s in readily accessible journals, I have sometimes given a double page reference in the following style: journal page no./book page no.]

– *The Royal Society and its Fellows 1660–1700. The morphology of an early scientific institution* (British Society for the History of Science Monographs, 1982)

Hunter, Michael and Wood, Paul B., 'Towards Solomon's house: rival strategies for reforming the Royal Society', *History of Science*, **24** (1986), 49–108 [reprinted in Hunter, *Establishing the New Science*, pp. 185–244]

Huygens, Christiaan, *Oeuvres Complètes* (The Hague, 1888–1950)

Jacquot, Jean, 'Sir Hans Sloane and French men of science', *Notes & Records*, **10** (1952), 85–98

Kaye, I., 'Unrecorded early meetings of the Royal Society', *Notes & Records*, **8** (1951), 150–2

Latham, R. and Matthews, W. (eds.), *The Diary of Samuel Pepys*, 11 vols. (London, 1970–83)

Leeuwenhoek, Antoni van, *Alle de Brieven* (Amsterdam, 1939–)

Lower, Richard, *De Corde* (London, 1669) [facsimile in Gunther (ed.), *Early Science in Oxford*, vol. IX (1932), with English translation by K. J. Franklin]

Lyons, H. G., 'Biographical notes. Francis Aston (1645–1715)', *Notes & Records*, **3** (1940), 88–92

– 'The Officers of the Society 1662–1860', *Notes & Records*, **3** (1940–4), 116–40

– *The Royal Society 1660–1940. A history of its Administration under its Charters* (Cambridge, 1944)

MacPike, E. F., *Correspondence and Papers of Edmond Halley* (Oxford, 1932; London, 1937)

Malpighi, *see* Adelmann

Maehle, Andreas-Holger, 'Literary responses to animal experimentation in 17th and 18th century Britain', *Medical History* **34** (1990), 27–51.

Mariotte, Savant et Philosophe (Paris, 1986)

Mélanges Alexandre Koyré, 2 vols. (Paris, 1964), I L'aventure de la science, introduction by I. B. Cohen and René Taton; II L'aventure de l'esprit, introduction by Fernand Braudel. Nos. 12 and 13 of *Histoire de la pensée*, Ecole pratique des Hautes Etudes

Middleton, W. E. K., 'What did Charles II call the Fellows of the Royal Society?', *Notes & Records*, **32** (1977), 13–16

Miller, D. P., ' "Into the Valley of Darkness"; reflections on the Royal Society in the eighteenth century', *History of Science*, **27** (1989), 155–66

Oldenburg, Henry, *et al.*, *Philosophical Transactions* (London, Oxford, 1665–1752) [thereafter officially a publication of the Royal Society]

Oldroyd, D. R., 'Some writings of Robert Hooke on procedures for the prosecution of scientific inquiry', *Notes & Records*, **41** (1987), 145–67

Ornstein, Martha, *The Rôle of Scientific Societies in the Seventeenth Century* (1913; 3rd edn, Chicago, 1938)

Oxford Philosophical Society, R. T. Gunther, 'Dr Plot and the correspondence of the Philosophical Society of Oxford', *Early Science in Oxford*, vol. XII (Oxford, 1939)

– 'The Philosophical Society of Oxford', *Early Science in Oxford*, vol. IV (Oxford, 1925)

Power, Henry, *Experimental Philosophy* (London, 1664) (facsimile edn. ed. M. B. Hall, New York and London, 1966)

Purver, Marjorie, *The Royal Society: Concept and Creation* (London, 1967)

Rappaport, Rhoda, 'Hooke on earthquakes: lectures, strategy and audience', *BJHS*, **19** (1986), 129–46

Rattansi, P. M., 'The intellectual origins of the Royal Society', *Notes & Records*, **23** (1968), 129–43

Raven, C. E., *John Ray, Naturalist* (Cambridge, 1950)

The Record of the Royal Society, 4th edn. (London, 1940)

Robinson, H. W., 'The administrative staff of the Royal Society (1663–1861)', *Notes & Records*, ...4, (1946), 193–205

Robinson, H. W. and Adams, W. (eds.), *The Diary of Robert Hooke M.D., F.R.S. 1672–1680* (London, 1935)

Roemer et la Vitesse de la Lumière (Paris, 1978)

Scala, G. E., 'An index of proper names in Thomas Birch, *The History of the Royal Society* (London, 1756–1757)', *Notes & Records*, **28** (1974), 263–329

Schaffer, Simon, 'Glass works: Newton's prisms and the uses of experiment', D. Gooding, T. Pinch and S. Schaffer (eds.), *The Uses of Experiment. Studies in the natural sciences* (Cambridge, 1989), pp. 67–104

– 'Wallifaction: Thomas Hobbes on school divinity and experimental pneumatics', *Studies in the History and Philosophy of Science*, **19** (1988), 275–98

Scriba, C., 'The autobiography of John Wallis, F.R.S.', *Notes & Records*, **25** (1970), 17–46

Shapin, Steven, 'The house of experiment in seventeenth century England', *Isis,* **79** (1988), 373–404

Shapin, Steven and Schaffer, Simon, *Leviathan and the Air Pump. Boyle and the experimental life* (Princeton, 1985)

Spiller, M.R.G., *'Concerning Natural Experimental Philosophie' Meric Causaubon and the Royal Society* (The Hague, Boston, London, 1980)

Sprat, Thomas, *The History of the Royal Society of London for the Improving of Natural Knowledge* (London, 1667). [Later, facsimile ed. Jackson I. Cope and Harold Whitmore Jones, St Louis, Missouri, 1958. All editions have the same pagination]

Stimson, Dorothy, 'Ballad of Gresham College', *Isis,* **18** (1932), 103–17

– *Scientists and Amateurs. A history of the Royal Society* (New York, 1948)

Syfret, R. H., 'Some early critics of the Royal Society', *Notes & Records,* **8** (1950), 20–64

Talon, Henri (ed.), *Selections from the Journals & Papers of John Byrom … (1691–1763)* (London, 1950)

Taylor, F. Sherwood, 'An early satirical poem on the Royal Society', *Notes & Records,* **5** (1947), 37–46

Thoresby, Ralph, *see* Hunter, Joseph

Turnbull, H. W., Scott, J. F., Hall, A. Rupert and Tilling, Laura, *The Correspondence of Isaac Newton,* 7 vols. (Cambridge, 1959–77)

Wallis, John, *A Defence of the Royal Society* (London, 1678)

Weld, C. R., *A History of the Royal Society,* 2 vols. (london, 1848)

Westfall, R. S., *Never at Rest: A biography of Isaac Newton* (Cambridge, 1980)

Wood, Paul B., 'Methodology and apologetic: Thomas Sprat's history of the royal society', *BJHS,* **13**, (1980), 1–26

Index

INDEX

Descartes, René (1596–1650), 52, 86
  followers of, 154, 155, 156, 157
Doody (Dowdy), Samuel (1656–1706), 19
Douglas, James (1675–1742), 127, 130
Dryden, John (1631–1700), 164, 188 n1
Dublin Philosophical Society, 84, 87, 88,
  89, 100
Du Quet, Mr, 127

Ent, George (d. 1679), 76
Evelyn, John (1620–1706), 13, 27 fig., 53,
  54, 67
  Fumifugium (1661), 54, 161
  Pomona (1664), 54
  Sculptura (1662), 54
  Sylva (1664), 54, 179 n4
  Terra (1676), 68

Fahrenheit, Daniel Gabriel (1686–1736),
  126
Fairchild, Thomas (1667?–1729), 127
Fairfax, Nathaniel (1637–90), 61
Fatio de Duillier, Nicolas (1664–1753),
  19, 89, 91, 96
Flamsteed, John (1646–1719), 63, 93, 100,
  114–15, 152, 153
Folkes, Martin (1690–1754), 124, 128
Franck von Frankenau, Georg
  (1643–1704), 150
Freind, John (1675–1728), 134

G., 109
Gale, Thomas (?1635–1702), 72 fig., 80,
  88, 101
Galiani, Celestino (1681–1753), 138, 146
Galilei, Galileo (1564–1642), 52
Gassendi, Pierre (1592–1655), 52
Geoffroy, Claude-Joseph (1685–1752),
  136, 137
Geoffroy, Etienne François (1672–1731),
  95–6, 105, 122, 134, 136, 137,
  153–4, 183 n31
Gilbert, W. S. (1836–1911), *Princess Ida*
  (1884), 167
Giornale de'Letterati, 62, 145, 146
Glanvill, Joseph (1636–80), 52–3, 56,
  163–6, 174 n3
  Plus Ultra (1668), 52
  Scepsis Scientifica (1664), 52
  Vanity of Dogmatizing (1661), 52
Goddard, Jonathan (1616–75), 15–16, 25,
  26, 34, 35, 64
  experiments of, 29, 30, 31, 33
Godfrey, Ambrose, 126, 186 n10
Godfrey, John, 96, 131, 186 n10, 187 n20
Graaf, Regnier De (1641–72), 147

Grandi, Guido (1671–1742), 138, 145
'sGravesande, Willem Jacob (1688–1742),
  134–5, 138, 147
  *Physices elementa mathematica ... introductio
  ad philosophiam Newtonianam* (1720),
  135
Gray, Stephen (1666–1736), 94, 95, 96,
  97, 126, 131
Greatorex, Ralph (1625–1712), 30
Gresham College, 25, 79
  see also Royal Society, locations of
Grew, Nehemiah (1641–1712), 19, 156
  as Curator, 44, 47, 48, 49, 178 n29
  and Hooke, 17, 18
  Musaeum Regalis Societatis (1681), 78
  papers read, 66, 68, 69, 70–78
  and *Phil. Trans.*, 27 fig., 107–8
  and plant anatomy, 44–5, 47
  as Secretary, 71, 72 fig., 73, 99, 100,
  101, 146
Grillet, Mr, 89, 90
Guericke, Otto von (1602–86), *Experimenta
  nova (ut vocantur) Magdeburgica*
  (1672), 47
Guglielmini, Domenico (1655–1710), 145

Haak, Theodore (1605–90), 17, 80, 101
Hadley, John (1682–1744), 127
Hale, Matthew (1609–76), *Difficiles nugae*
  (1674), 156
Hales, Stephen (1677–1761), 112, 125,
  134
  Vegetable Staticks (1727), 130, 134
  French ed. of (1735), 134
Hall, A. Rupert and Marie Boas, eds.,
  The Correspondence of Henry Oldenburg
  (1965–86), xi
Hall, Marie Boas, 'Salomon's house
  emergent: the Royal Society and
  cooperative research', 2–3
  'Science in the early Royal Society', 2–3
Hall, Marshall (1790–1857), 1–2, 132
Halley, Edmond (1656–1742), 89, 95, 96,
  127, 159
  as astronomer, 94, 114–15, 126
  as Clerk, 72 fig., 87, 89, 91, 101, 103,
  104, 107, 117 fig., 184 n13
  experiments by, 84, 85, 87–8, 89, 90,
  91, 122
  and Newton, 88, 103, 110
  papers by, 87, 89, 90, 91–2, 108, 126,
  127
  and *Phil. Trans.*, 91–5, 97, 110–11, 117
  fig., 120, 127, 135
Hamilton, Sir William (1730–1803), 137
Hanneman, Dr, 147

INDEX

Huygens, Constantijn (1596–1687), 147

Istituto delle Scienze, Bologna, 145

Jones, Jezreel (d. 1731), 104
Journal des Sçavans, 37, 60, 61, 62, 78, 89,
134, 151–2
Jurin, James (1684–1750), 117 fig., 125,
128, 129, 130, 135
Jussieu, Antoine de (1686–1758), 127
Justel, Henri (1620–93), 99, 101, 152,
182 n16

Keill, John (1671–1721), 134, 138
King, Edmond (1629–1709), 17, 36–7, 40,
68, 93, 156
King, William (1663–1712), 167–8
Transactioneer (1700), 168
Kirke, Thomas (1650–1706), 94
Kraft, J. D., (1624–97), 150
Kunckel, Johann (?1630–1702), *Chymischer
Probierstein de acido & urinoso . . .*
(1684), 85

Lavoisier, A. L. de (1743–94), 160,
190 n37
Leeuwenhoek, Antoni van (1632–1723),
61, 73, 99, 101–2, 146–7, 164,
183 n2
letters of, 70–1, 72, 74, 75, 80, 81, 89,
92–9 passim, 102, 108, 110, 112,
120, 121, 123, 125, 130, 183 n2
microscopes, 48, 73, 76–7
Leibniz, Gottfried Wilhelm (1646–1716),
57, 63, 99, 129, 133, 150–1, 153,
154
arithmetical machine, 48, 139, 149
Hypothesis physica nova (1671), 43, 149,
152
see also Newton, Isaac
L'Hopital, Guillaume François Antoine
de (1661–1704), 105
Lilly, William (1602–81), 162
Line or Linus (1595–75), 63, 68, 158
Lister, Martin (1639–1712), 61, 63, 85,
96, 99, 106, 115
his blood-staunching liquor, 49
Journey to Paris (1699), 167–8
on plant physiology, 44, 47, 95
visit to Paris, 105
Lodwick, Francis (1619–94), 18
Long, Captain, 74
Louville, Jacques-Eugène d'Allonville,
chevalier de (1671–1732), 136, 137
Lower, Richard (1631–91), 34, 35–8, 40,
43

Tractatus de Corde (1669), 35, 37, 43
Lowther, Sir John (1642–1706), 67
Lowthorp, John, (F.R.S. 1702), 96,
183 n32
Lucas, Anthony (1633–93), 99
Ludolf, Job (1624–1704), 148

Machin, John (1699?–1751), 117 fig.
Magalotti, Lorenzo (1637–1712), 143
Malpighi, Marcello (1628–94), 44–5, 57,
99, 101–2, 143
Anatomes plantarum pars alterum (1679),
106
books sponsored, 59, 98, 105–6, 143,
180 n16
'Dissertatio epistolica varii argumenti'
(1684), 106
Opera omnia (1686), 106
Opera posthuma (1697), 102, 106
De Structura glandulorum . . . epistola
(1689), 106
Mariotte, Edmé (d. 1684), 153, 154
Essai de logique (1678), 153
Marsigli, Luigi Ferdinando, Count
(1658–1730), 138, 145
Martyn, John, 97
Maty, Matthew (1718–76), 136
Merret, Christopher (1615–95), 29
Mersenne, Marin (1588–1648), 58
Millington, Thomas (1628–1704), 35
Miscellanea Curiosa, 148
Molesworth, Richard (1680–1758), 129
Molyneux, William (1656–98), 100
Montague, Charles (1661–1715), 72 fig.
Montanari, Geminiano (1637–87), 144,
145
Montmort, Pierre Rémond de
(1678–1719), 136–7
Moore, Jonas (1617–79), 17, 155
Moray, Sir Robert (1608–73), 29, 31, 34,
35, 39, 45, 56, 59, 146, 161
and experiments, 30, 34
Mortimer, Cromwell (b. 1752), 136
Moulin or Mullen, Allen (c. 1653–c. 90),
89, 90
Moult, George (d. 1727), 19, 96
Munchausen, Benjamin von, *De Herniis*,
173 n9
Musgrave, William (?1655–1721), 72 fig.,
100, 110
Musschenbroek, Petrus van (1692–1761),
147

Nazari, Francesco (1634–1714), 62, 145
Needham, Walter (?1631–91), 16, 68
Neile, Sir Paul (c. 1613–86), 14, 45, 161

INDEX

Poleni, Giovanni (1683–1761), 137, 145
Pope, Alexander (1688–1744), 167, 168
  Essay on Man (1733), 166
Pope, Walter (*c.* 1627–1714), 38
Povey, Thomas (b.*c.* 1615), 30, 93
Power, Henry (1623–68), 30, 31, 54–5
  Experimental Philosophy (1664), 30, 55
Pryme, Abraham de la (1672–1704), 97

Ranelagh, Lady Katherine (1614–91), 76
Ray, John (1627–1705), 66, 106–7, 115,
  162
Riccati, Jacopo Francisco (1675–1754), 146
Richardson, Richard (1663–1741), 97
Rizzetti, Giovanni, 146
Robinson, Richard (d. 1733), 19
Robinson, Tancred (*c.* 1657–1748), ?19,
  72 fig., 103, 106, 185 n22
Roemer, Olaus (1644–1710), 153
Rohault, Jacques, *Traité de physique* (1671),
  134
Rooke, Lawrence (1622–62), 9, 25, 26, 30
Royal Society
  and Académie Royale des Sciences,
    137, 154
  activities at meetings, xi, 1, 2, 6–7, 15,
    22, 24–49, 66–97, 116–32, 140
  aims of, 4–5, 9–23, 50ff, 57, 58, 111,
    112, 113, 120, 142, 159
  amanuensis, 26, 53, 68, 79
    see also Wicks, Michael
  attendance at meetings, 17
  as Baconian, 9–10, 52, 151
  beginnings of, 9, 24, 176 n2
  calendar of, xiii
  charters of, 10, 31, 53, 54, 64
  clerks, 72 fig.; *see also* Halley *and* Wicks
  committees of, 1, 14, 15, 20, 27, 29, 36,
    77, 175 n21
  correspondence of, 54–8, 100–105; *see*
    also under individual Secretaries
  Curators of Experiment, 5, 7, 13, 14,
    22, 31, 36, 38, 39, 43, 47–8, 71, 72
    fig., 73, 77, 79, 80, 83, 85, 88,
    116 ff, 123, 131; *see also*
    Desaguliers, Hauksbee, Hooke,
    Papin, Slare, Tyson
  discussion at meetings, 4, 131
  empiricism of, criticised, 155, 169
  empiricism of, praised, 140–54 passim
  and histories of trades, 10
  hypotheses, attitude to, 10–11, 147
  institutional history of, 3
  Journal Books, xi, 3, 4, 53, 63, 176 n1
  lectures at meetings, 15–16, 19, 48,
    66–7, 67 fig., 122 fig.

Letter Books, 70, 99, 176 n1
library, 14
locations of, 13–14
  Arundel House, 13, 38
  Gresham College, 13, 38, 49, 97
and mathematics, 10
'openness' of, 55–6
and *Phil. Trans.*, 107–13, 120, 131, 136
proposals for reform of, 3, 14–23, 39,
  40, 43, 49, 64, 79–80, 174 n12
Register Books, 28, 41, 67, 68, 80, 85,
  107
repository, 78
reputation, 5, 140–69
satires on, 161–7
secrecy, 64–6
Secretaries, 98–9, 101, 108, 131, 142; *see*
  also figs 1, 5, 6
sponsorship of publication, 44–5,
  105–7, 115, 143, 180 n16
statutes, 1, 65, 71, 187 n21
'strangers' at meetings, 65, 71, 94
and universal natural history, 10, 40
Rupert, Prince (1619–82), 64, 80, 179 n10
Rutty, William (1687–1730), 117 fig., 135

Sand, Christoph (1644–80), 190 n19
Schaffer, Simon, xii, 176 n24
Savery's steam engine, 96, 125
Schullian, Dorothy, xii
Seaforth, Mr, 96
Shadwell, Thomas (1642?–92), 164, 167,
  191 n43
  The Virtuoso (1676), 164–5
Shaen, Sir James (d. 1695), 67
Shapin, Steven, 'The house of experiment
  in seventeenth century England', 3
  and Schaffer, *Leviathan and the Air Pump*,
  3, 190 n35
Shaw, Peter (1694–1764), 139
  Chemical Lectures (1734), 135
  A New Method of Chemistry (1727), 134
Shortgrave, Richard (d. 1676), 27 and 27
  fig.
Silvestre (?Sister), Pierre (?1662–1718),
  19
Sinclair, George (d. 1696), *Ars nova &*
  magna gravitatis et levitatis (1669), 156
Skippon, Philip (1641–91), 162
Slare, Frederick (1648–1727), viii, 78, 85,
  86, 91, 95, 110, 114, 150, 188 n4
  and Boyle, 83, 86, 114, 118
  as Curator, 72 fig., 83, 84, 85, 86
  experiments of, 82, 83, 85, 90, 93, 96, 141
Sloane, Hans (1660–1753), 19, 91, 103,
  124, 126, 128, 130, 185 n23

~ 205 ~

INDEX

Wren, Christopher (1632–1723), 1,
  14, 17, 18, 24, 25, 27, 34, 39–40,
  45, 49, 71, 74, 152, 190 n41,
  191 n42
  on motion, 14, 41–2
  performing experiments, 30
  and physiology, 35

as President, 71, 72 fig., 80, 101, 117
  and Wallis, 114
Wyche, Cyril (?1632–1707), 72 fig.
Wylde, Edmund (c. 1614–96), 17

Yonge, James (1646–1721), 108

Zanotti, F. M. (1692–1777), 145

Printed in the United States
By Bookmasters